**Topological
vector spaces**

Notes on mathematics and its applications

General editors: Jacob T. Schwartz, *Courant Institute of Mathematical Sciences,* and Maurice Lévy, *Université de Paris*

E. Artin ALGEBRAIC NUMBERS AND ALGEBRAIC FUNCTIONS
R. P. Boas COLLECTED WORKS OF HIDEHIKO YAMABE
R. A. Bonic LINEAR FUNCTIONAL ANALYSIS
R. B. Burckel WEAKLY ALMOST PERIODIC FUNCTIONS ON SEMIGROUPS
M. Davis A FIRST COURSE IN FUNCTIONAL ANALYSIS
M. Davis LECTURES ON MODERN MATHEMATICS
J. Eells, Jr. SINGULARITIES OF SMOOTH MAPS
K. O. Friedrichs ADVANCED ORDINARY DIFFERENTIAL EQUATIONS
K. O. Friedrichs SPECIAL TOPICS IN FLUID DYNAMICS
A. Grothendieck TOPOLOGICAL VECTOR SPACES
A. Guichardet SPECIAL TOPICS IN TOPOLOGICAL ALGEBRAS
M. Hausner and J. T. Schwartz LIE GROUPS; LIE ALGEBRAS
P. Hilton HOMOTOPY THEORY AND DUALITY
S. Y. Husseini THE TOPOLOGY OF CLASSICAL GROUPS AND RELATED TOPICS
F. John LECTURES ON ADVANCED NUMERICAL ANALYSIS
A. M. Krall STABILITY TECHNIQUES FOR CONTINUOUS LINEAR SYSTEMS
R. L. Kruse and D. T. Price NILPOTENT RINGS
P. Lelong PLURISUBHARMONIC FUNCTIONS AND POSITIVE DIFFERENTIAL FORMS
H. Mullish AN INTRODUCTION TO COMPUTER PROGRAMMING
D. Ponasse MATHEMATICAL LOGIC
F. Rellich PERTURBATION THEORY OF EIGENVALUE PROBLEMS
J. T. Schwartz DIFFERENTIAL GEOMETRY AND TOPOLOGY
J. T. Schwartz NON-LINEAR FUNCTIONAL ANALYSIS
J. T. Schwartz W-* ALGEBRAS
M. Shinbrot LECTURES ON FLUID MECHANICS
G. Sorani AN INTRODUCTION TO REAL AND COMPLEX MANIFOLDS
J. L. Soulé LINEAR OPERATORS IN HILBERT SPACE
J. J. Stoker NONLINEAR ELASTICITY
F. Trèves LINEAR PARTIAL DIFFERENTIAL EQUATIONS
F. Trèves LINEAR PARTIAL DIFFERENTIAL EQUATIONS WITH CONSTANT COEFFICIENTS

Additional volumes in preparation

Topological
vector spaces

A. GROTHENDIECK

Institut des Hautes Études Mathématiques
Bures-sur-Yvette

Translated from the French by
ORLANDO CHALJUB

GORDON AND BREACH
Science Publishers
NEW YORK   LONDON   PARIS

Library of Congress catalog card number 72–83698. ISBN 0 677 30020 4 (cloth); 0 677 30025 5 (paper). All rights reserved. No part of this book may be reproduced or utilized in any form or by any means, electronic or mechanical, including photocopying, recording, or by any information storage or retrieval system, without permission in writing from the publishers. Printed in Great Britain.

Foreword

The present book consists of a fairly literal translation of a course in
French which I gave in São Paulo in 1954, and which has been available
since then in only a mimeographed form. For particulars of the material
covered in this course, the reader should consult the Introduction
(page 1) as well as the Contents (page vii). The author has not revised
this account of the theory of topological vector spaces, first given
fourteen years ago, mainly because he has not been directly involved
with the subject for nearly as many years, and because other tasks
(including some of a similar nature) have kept him rather too busy.
This, the author feels, is of not too serious consequence, since the
material covered in his São Paulo course has remained practically
unchanged since that time, so that the revision would have resulted in
scarcely more than minor stylistical changes, which any œsthetically
minded reader would be able to provide by himself. Generally speaking,
it seems that the theory of locally convex spaces has not significantly
progressed since that time, most probably because no such progress was
necessary in related fields.

Bibliographical data are not included in this book, since the reader
will be able to find such information in any of the various books on the
same subject which have recently appeared, among which are the books
by Bourbaki, and by Koethe (*Topologische Lineare Räume I*, Springer-
Verlag 1960), the latter also containing numerous results not given in
the present work.

Finally, I am glad to extend my warmest thanks to O. Chaljub for
his careful translation of the original French into English, and also for
his help in proofreading. Equal thanks are due to Mr E. Thomas for
reading the whole manuscript critically, correcting various mistakes,
and making several valuable terminological suggestions for the English
translation.

A. GROTHENDIECK

Contents

Topological introduction

As a prerequisite for this course, the reader should know some general topology as given in *Topologie Générale* by Bourbaki, particularly Chapters I, II, IV and IX, Chapter VI, Sections 1 and 2 (topological properties of $\mathbf{R}^n$), Chapter VIII (complex numbers) and, in particular, Chapter III (topological groups). Non-abelian topological groups will not be used. From the book *Algebra* by Bourbaki, we shall use Chapters I and II; notice that we shall always suppose that the underlying field is the real field $\mathbf{R}$ or the complex field $\mathbf{C}$.

We wish to recall here some points of general topology which will be of particular interest to us and, still more particularly, a part of Chapter X of Bourbaki's *Topologie Générale*.

Most proofs, requiring only simple verifications, are left to the reader.

1 Least upper bound of a family of topologies

Consider a non-empty family $(T_i)_{i \in I}$ of topologies on a set E. We know that a *least upper bound topology* of the topologies T_i exists, i.e. the coarsest topology on E, finer than each T_i. If $\mathscr{V}_i$ is the set of open sets for the topology T_i $(i \in I)$, the least upper bound topology is generated by the union of the $\mathscr{V}_i$.

Let E be any set and $(E_i)_{i \in I}$ a non-empty family of topological spaces and, for each $i \in I, f_i$ a mapping of E into E_i $(i \in I)$. There is a topology called the *initial topology* of the E_i by the mappings f_i, which is the coarsest topology for which all the mappings f_i are continuous: it is the least upper bound of those topologies which are inverse images of the topologies of the space E_i, by the mappings f_i.

Separation condition In order that the space E, equipped with the initial topology of the E_i induced by the f_i, be Hausdorff, it is necessary, and also sufficient, if the E_i are Hausdorff, that for $x \neq y$ in E there exists an f_i such that $f_i(x) \neq f_i(y)$.

Transitivity of the initial topology If the topology of E is the initial topology of the E_i by the f_i and if the topology of each E_i is also the initial topology of spaces E_{ik} induced by mappings ϕ_{ik} of E_i into E_{ik}

($k \in A_i$), then the topology of E is the initial topology of the E_{ik} by the mappings $\phi_{ik} \circ f_i$ ($i \in I$, $k \in A_i$).

In particular, if F is a subset of E, the topology induced on F by the topology of E is the initial topology of the E_i by the restrictions of the f_i to F. Another particular case: the initial topology of the E_i by the f_i is the inverse image of the topology of the product space $\prod_{i \in I} E_i$ by the mapping

$$x \longrightarrow f(x) = (f_i(x))_{i \in I}$$

from E into $\prod_{i \in I} E_i$; if for any x, $y \in E$, $x \neq y$, there exists an $i \in I$ such that $f_i(x) \neq f_i(y)$, then f is an isomorphism of E onto a subspace of the product $\prod_{i \in I} E_i$. (Notice that by definition, the topology of $\prod_{i \in I} E_i$ is the initial topology of the E_i by the natural projections $E \rightarrow E_i$.)

PROPOSITION 1   *Let E be a topological space with the initial topology of spaces E_i by mappings f_i.*

1) Let ϕ be a mapping of a topological space F into E. The mapping ϕ is continuous if and only if for every i the mappings $f_i \circ \phi$ from F into E_i is continuous.

2) A filter Φ on E converges to $x \in E$ if and only if for every i, the filter base $f_i(\Phi)$ on E_i converges to $f_i(x)$.

If E is a set, $(E_i)_{i \in I}$ a non-empty family of topological spaces, and $f_i(i \in I)$ mappings from E_i into E, we can consider the finest topology on E for which the f_i are continuous. The open (or closed) sets in this topology are the subsets U of E such that the $f_i^{-1}(U)$ are open (or closed) in E_i for every $i \in I$. This topology is called the *final topology* of the E_i induced by the f_i. Let f be a mapping from E into a topological space F, then f is continuous if and only if each $f \circ f_i$ is continuous.

A particular case   Given a family $(A_i)_{i \in I}$ of subsets of E, each A_i with a topology T_i. Then, there exists on E a topology T, the finest among those which induce a topology coarser than the T_i on the sets A_i. The open (or closed) subsets for T are those whose intersection with each A_i is open (or closed) for T_i. A mapping f from E into a topological space is continuous if and only if its restriction to every A_i is continuous for T_i. If we can find on E a topology T' that induces on every A_i exactly T_i, then $T \geqslant T_i$, therefore T induces T_i on every A_i The most important case of the situation just described will be seen in Chapter 4, Part 2, Section 3, Theorem 2.

2 Least upper bound of a family of uniform structures

Consider on a set E a non-empty family $(\mathscr{U}_i)_{i \in I}$ of uniform structures. We know there exists a *l.u.b.* uniform structure for the $\mathscr{U}_i$, i.e., the coarsest uniform structure on E among those which are finer than each $\mathscr{U}_i$. The entourage filter of this uniform structure in the filter generated by the union of the entourage filters of the $\mathscr{U}_i$. The set of intersections of a finite number of entourages W_i in $\mathscr{U}_i (1 \leqslant i \leqslant n)$ is a fundamental system of entourages of $\mathscr{U}$. *The topology associated with $\mathscr{U}$ is the l.u.b. topology for the topologies associated with the $\mathscr{U}_i$.*

If E is a set, $(E_i)_{i \in I}$ a non-empty family of uniform spaces, and if for each $i \in I$, f_i is a mapping from E into E_i, then there exists on E a uniform structure called the *initial uniform structure* of the E_i induced by the mappings f_i, which is the coarsest for which the f_i are uniformly continuous. It is the l.u.b. of the uniform structures which are the inverse images by the f_i of the uniform structures on E_i. The topology deduced from this uniform structure is the finest on E for which the f_i are continuous. Conversely, the uniform structure $\mathscr{U}$, l.u.b. of a family $(\mathscr{U}_i)_{i \in I}$ of uniform structures, can be considered as the coarsest for which the identity mappings $E \to E_i$ are uniformly continuous (E_i stands for E equipped with the uniform structure $\mathscr{U}_i$).

Separation condition The space E which has the initial uniform structure of the E_i induced by the f_i is Hausdorff if and only if for $x, y \in E$, $x \neq y$, there exists an index i and an entourage V_i of E_i such that $(f_i(x), f_i(y)) \notin V_i$. Furthermore, if the E_i are *Hausdorff* uniform spaces, it is necessary and sufficient that for $x, y \in E$, $x \neq y$, there exists an index i such that $f_i(x) \neq f_i(y)$.

Transitivity of the initial uniform structure If we substitute "uniform structure" for "topology" and "uniform space" for "topological space", we can, word for word, repeat the section on transitivity of the initial topology.

PROPOSITION 2 *Let E be a uniform space with the initial uniform structure of the E_i by the f_i.*

1) *A filter Φ on E is a Cauchy filter if and only if for every $i \in I$, $f_i(\Phi)$ is the base of a Cauchy filter on E_i.*

2) *Let ϕ be a mapping from a uniform space F into E. The mapping ϕ is uniformly continuous if and only if for every $i \in I$, the mapping $f_i \circ \phi$ from F into E_i is uniformly continuous.*

3 Precompact spaces

A uniform space E is called *precompact* if the completion of the *Hausdorff* space associated with E is compact. (Contrary to Bourbaki, a precompact space is not necessarily Hausdorff.) A subset A of E is precompact if the uniform subspace A is precompact. The closure of A in E is then also precompact.

PROPOSITION 3

1) *A uniform space E is precompact if and only if, for every entourage V of E, there exists a finite covering of E by V-small sets.*

2) *Let f be a uniformly continuous mapping from a precompact space E into a uniform space F; then $f(E)$ is a precompact subset of F.*

3) *Let E be a space equipped with the coarsest uniform structure for which the mappings f_i into the uniform spaces E_i are uniformly continuous. Then E is precompact if and only if for every i $f_i(E)$ is precompact.*

From 3) we conclude also that a subset A of E is precompact if and only if for every $i \in I$, $f_i(A)$ is a precompact subset of E_i.

EXERCISE   A space E is precompact if and only if every ultrafilter on E is a Cauchy filter.

4 $\mathfrak{S}$-convergence

Let E be any set and let F be a uniform space. Write $\mathscr{F}(E, F)$ for the space of all mappings from E into F. Let A be a subset of E, U an entourage of F and $W(A, U)$ the set of pairs (u, v) of mappings of E into F such that $(u(x), v(x)) \in U$ for every $x \in A$. If we keep A fixed and let U run through the entourage filter of F (or a fundamental system of entourages of F), the sets $W(A, U)$ form a fundamental system of entourages for a uniform structure on $\mathscr{F}(E, F)$, called *the uniformity of uniform convergence on A* or the A-convergence uniform structure. A filter Φ converging towards u_0 for the topology deduced from this uniform structure is said to be uniformly convergent to u_0 *in A*.

Let $\mathfrak{S}$ be a set of subsets of E. The l.u.b. of the A-convergence uniform structures ($A \in \mathfrak{S}$) is called the $\mathfrak{S}$-*convergence uniform structure*. We denote by $\mathscr{F}_{\mathfrak{S}}(E, F)$ the set $\mathscr{F}(E, F)$ with this uniform structure. We have a fundamental system of entourages if we choose for each entourage U of the entourage filter of F (or of a fundamental entourage system of F) and for each *finite* sequence $(A_i)_{1 < i < n}$ of elements of $\mathfrak{S}$, the entourage

$$W\left(\bigcup_{i=1}^{n} A_i, U\right) = \bigcap_{i=1}^{n} W(A_i, U).$$

It is immediate that the $\mathfrak{S}$-convergence uniform structure is the coarsest uniform structure on $\mathscr{F}(E, F)$ for which every mapping $\rho_A (A \in \mathfrak{S})$ from $\mathscr{F}(E, F)$ into the $\mathscr{F}(A, F)$ equipped with A-convergence is continuous, $\rho_A(u)(u \in \mathscr{F}(E, F))$ being the *restriction* of u to A.

The $\mathfrak{S}$-convergence uniform structure does not change if $\mathfrak{S}$ is replaced by the set $\mathfrak{S}'$ of *finite* unions of sets of $\mathfrak{S}$, or by the set $\mathfrak{S}''$ of all subsets of members of $\mathfrak{S}'$. If $\mathfrak{S} = \mathfrak{S}'$ we have a fundamental entourage system by choosing the sets $W(A, U)$ for every $A \in \mathfrak{S}$ and for every U of the entourage filter of F (or of a fundamental entourage system of F).

EXAMPLES

1) Choose $\mathfrak{S}$ to be the set of one-point subsets of E, or, what is equivalent, the set of finite subsets of E. The $\mathfrak{S}$-convergence uniform structure we obtain on $\mathscr{F}(E, F)$ is called *the uniform structure of simple (or pointwise) convergence*; we denote by $\mathscr{F}_s(E, F)$ the set $\mathscr{F}(E, F)$ with this uniform structure. If for every $x \in E$ we denote by $\tilde{x}$ the mapping from $\mathscr{F}(E, F)$ into F which to each $u \in \mathscr{F}(E, F)$ assigns $u(x)$ (that is, $\tilde{x}(u) = u(x)$), then the simple convergence uniform structure is the coarsest uniform structure for which the $\tilde{x}$ are uniformly continuous; equivalently, $\mathscr{F}(E, F)$ is the *product* space F^E.

If a filter Φ on $\mathscr{F}(E, F)$ converges to u_0 for the associated topology we say that Φ *converges simply* towards u_0. If $E_0 \subset E$, we shall call $\mathfrak{S}$-convergent uniform structure on $\mathscr{F}(E, F)$, where $\mathfrak{S}$ is the set of finite subsets of E_0, *the uniformity of simple convergence in E_0*.

2) If $\mathfrak{S} = \{E\}$, the uniform structure we obtain is called *the uniformity of uniform convergence*. We will call the set $\mathscr{F}(E, F)$ with this structure $F_u(E, F)$. If a filter Φ converges towards u_0 for the corresponding topology we say that Φ *converges uniformly towards u_0 in E*.

3) Let E be a topological space, let $\mathfrak{S}$ be the set of all *compact* subsets of E. The corresponding uniform structure is called the *uniform structure of compact convergence*; we will call the set $\mathscr{F}(E, F)$ with this uniform structure $\mathscr{F}_c(E, F)$. If a filter Φ converges towards u_0 for the corresponding topology then we say that it *converges uniformly towards u_0 in every compact set*.

E being a topological space, the simple convergence uniform structure is coarser than the compact convergence uniform structure which is coarser than the uniformly convergence uniform structure.

PROPOSITION 4 *Let $\mathfrak{S}$ be a set of subsets of E, F a uniform space.*

1) *A filter Φ on $\mathscr{F}_{\mathfrak{S}}(E, F)$ converges to u_0 if and only if for every $A \in \mathfrak{S}$, Φ converges to u_0 uniformly in A*

2) Φ *is a Cauchy filter if and only if for every* $A \in \mathfrak{S}$, Φ *is a Cauchy filter for A-convergence.*

3) $M \subset \mathscr{F}_\mathfrak{S}(E, F)$ *is precompact if and only if for every* $A \in \mathfrak{S}$, M *is precompact for A-convergence.*

4) *A mapping f from a topological (or uniform) space H into* $\mathscr{F}_\mathfrak{S}(E, F)$ *is continuous (or uniformly continuous) if and only if for every* $A \in \mathfrak{S}$, *f is continuous (or uniformly continuous) for A-convergence.*

The proof follows immediately from the definition of $\mathfrak{S}$-convergence and from Propositions 1, 2, 3.

REMARK   If we consider $\mathfrak{S}$-convergence as the coarsest uniform structure on $\mathscr{F}(E, F)$ for which the restrictions $\rho_A (A \in \mathfrak{S})$ from $\mathscr{F}(E, F)$ into the $\mathscr{F}_u(A, F)$ are uniformly continuous, then, by Propositions 1, 2 and 3 we obtain variants for the criteria of Proposition 4. Thus we can rewrite 1): a filter Φ on $\mathscr{F}_\mathfrak{S}(E, F)$ converges towards u_0 if and only if for every $A \in \mathfrak{S}$ the filter base $\Phi_A = \rho_A(\Phi)$ (formed by the restrictions of the $u \in \Phi$ to A) converges uniformly to $\rho_A(u_0)$ (the restriction of u_0 to A). Similar variants can be given for Criteria 2, 3 and 4 of Proposition 4.

PROPOSITION 5   *The uniform space* $\mathscr{F}_\mathfrak{S}(E, F)$ *is Hausdorff if and only if F is Hausdorff and* $E = \bigcup_{A \in \mathfrak{S}} A$.

In order to see that $\mathscr{F}_\mathfrak{S}(E, F)$ Hausdorff implies F Hausdorff we notice that F is isomorphic to the subspace of $\mathscr{F}_\mathfrak{S}$ formed of constant mappings from E into F.

THEOREM 1 $\mathscr{F}_\mathfrak{S}(E, F)$ *is complete if and only if F is complete.*

The necessity follows if we consider F as the subspace of constant mappings from E into F. The sufficiency follows from

PROPOSITION 6   *Let* Φ *be a filter on* $\mathscr{F}_\mathfrak{S}(E, F)$. *Then* Φ *converges to* u_0 *if and only if* Φ *is a Cauchy filter for* $\mathfrak{S}$-*convergence and* Φ *converges to* u_0 *for the uniformity of simple convergence in* $E_0 = \bigcup_{A \in \mathfrak{S}} A$ *(that is, for every* $x \in E_0$, $\Phi(x)$ *converges to* $u_0(x)$ *in F).*

The verification is immediate. We conclude:

COROLLARY   *Let* $\mathfrak{S}_1$ *and* $\mathfrak{S}_2$ *be sets of subsets of E such that* $\mathfrak{S}_1 \subset \mathfrak{S}_2$ *and*

$$\bigcup_{A \in \mathfrak{S}_1} A = \bigcup_{B \in \mathfrak{S}_2} B;$$

let H be a subset of $\mathscr{F}(E, F)$.

1) *If H is complete for $\mathfrak{S}_1$-convergence then H is complete for $\mathfrak{S}_2$-convergence.*

2) *If H is precompact for $\mathfrak{S}_2$-convergence, then on H, the $\mathfrak{S}_1$-convergence and the $\mathfrak{S}_2$ convergence uniform structures coincide.*

1) follows from Proposition 6. In order to show 2) we shall suppose $\bigcup_{B \in \mathfrak{S}_2} B = E$ and F Hausdorff for the sake of simplicity. We can suppose F complete (if not, use the completion) and H closed in $\mathscr{F}_{\mathfrak{S}_2}(E, F)$. Since this space is Hausdorff, H is even compact, therefore its uniform structure is identical to every coarser *Hausdorff* uniform structure, in particular to the $_{A_1}$-convergence structure. In fact, it is not necessary that $E = \bigcup_{B \in \mathfrak{S}_2} B$ and F be Hausdorff, as we shall see from Exercise 1 below.

PROPOSITION 6′ *The set of $u \in \mathscr{F}_{\mathfrak{S}}(E, F)$ which transform the $A \in \mathfrak{S}$ into precompact subsets of F is closed.*

This is a simple consequence of the usual criteria of precompactness (Proposition 3, 1).

COROLLARY *If F is Hausdorff and complete, the space of mappings from E into F which transform the $A \in \mathfrak{S}$ into relatively compact subsets is complete for $\mathfrak{S}$-convergence.*

$\mathfrak{S} \times \mathfrak{T}$ — *convergence on a product*

PROPOSITION 6″ *Let E, F be sets, $\mathfrak{S}$ a set of subsets of E, $\mathfrak{T}$ a set of subsets of F, $\mathfrak{S} \times \mathfrak{T}$ the set of subsets of $E \times F$ of the form $A \times B$ ($A \in \mathfrak{S}$, $B \in \mathfrak{T}$), G a uniform space.*
Then the uniform spaces: $\mathscr{F}_{\mathfrak{S} \times \mathfrak{T}}(E \times F, G)$ and $\mathscr{F}_{\mathfrak{S}}(E, \mathscr{F}_{\mathfrak{T}}(F, G))$ are isomorphic (by the canonical mapping from the first onto the second defined in set theory).

This follows from the definitions of the respective entourages.

EXERCISES

1) The Hausdorff space associated with $\mathscr{F}_{\mathfrak{S}}(E, F)$ can be identified with $\mathscr{F}_{\mathfrak{S}}(E_0, F_0)$ where $E_0 = \bigcup_{A \in \mathfrak{S}} A$ and F_0 is the Hausdorff space associated with F.

2) If $\mathfrak{S}_1$ and $\mathfrak{S}_2$ are two sets of subsets of E such that on $\mathscr{F}(E, F)$ the $\mathfrak{S}_1$-convergent uniform structure coincides with the $\mathfrak{S}_2$-convergent uniform structure, then $\mathfrak{S}_1″ = \mathfrak{S}_2″$.

B

5 $\mathfrak{S}$-convergence in the spaces of continuous mappings

In general, we should consider several important subsets of the space of *all* mappings from E into F. Let E be a *topological space* and F a uniform space; we denote by $\mathscr{C}(E, F)$ the subset of $\mathscr{F}(E, F)$ formed by the continuous mappings. If we have $\mathfrak{S}$-convergence on $\mathscr{F}(E, F)$ we write $\mathscr{C}_{\mathfrak{S}}(E, F)$ for $\mathscr{C}(E, F)$ with the induced uniform structure. We shall write $\mathscr{C}_s(E, F)$, $\mathscr{C}_u(E, F)$ and $\mathscr{C}_c(E, F)$ for the set $\mathscr{C}(E, F)$ with the simply, uniformly and compact convergence uniform structures respectively.

Since the $u \in \mathscr{C}(E, F)$ are continuous we can easily see that if we replace $\mathfrak{S}$ by the set of closures of the $A \in \mathfrak{S}$, then $\mathscr{C}_{\mathfrak{S}}(E, F)$ does not change.

PROPOSITION 7   *For the space $\mathscr{C}_{\mathfrak{S}}(E, F)$ to be Hausdorff it is necessary that F be Hausdorff and this condition is sufficient if $E_0 = \bigcup_{A \in \mathfrak{S}} A$ is dense in E.*

THEOREM 2   *The set $\mathscr{C}(E, F)$ is closed in $\mathscr{F}_u(E, F)$.*
   That is: every uniform limit of continuous functions is continuous.

COROLLARY 1   *$\mathscr{C}_u(E, F)$ is complete if and only if F is complete.*

In order to show the sufficiency use Theorems 1 and 2; for the necessity proceed as for Theorem 1.

COROLLARY 2   *The space $\mathscr{C}_{\mathfrak{S}}(E, F)$ is closed in $\mathscr{F}_{\mathfrak{S}}(E, F)$ whenever each mapping from E into F whose restrictions to the $A \in \mathfrak{S}$ are continuous, is already continuous. In this case $\mathscr{C}_{\mathfrak{S}}(E, F)$ is complete if and only if F is complete.*

The hypothesis of the corollary is satisfied, for example, when E is locally compact or metrisable, and $\mathfrak{S}$ is the set of compact subsets of E.

PROPOSITION 8   *The mapping $(u, x) \mapsto u(x)$ from $\mathscr{C}_u(E, F) \times E$ into F is continuous.*

6 Equicontinuous and uniformly equicontinuous sets

Let E be a topological space, F a uniform space and H a subset of $\mathscr{F}(E, F)$. We say that H is *equicontinuous at the point x_0* of E if for every entourage U of F there exists a neighbourhood V of x_0 in E such that $(u(x), u(x_0)) \in U$ for every $x \in V$ and every $u \in H$. The set H is *equi-*

continuous in E if *H* is equicontinuous at each point of *E*. Then clearly each *u* ∈ *H* is continuous at x_0 (or continuous in *E*). In the case where *E* is a uniform space we say that *H* is uniformly equicontinuous if for every entourage *U* in *F* there exists an entourage *V* in *E* such that $(u(x), u(y)) \in U$ for every $(x, y) \in V$ and every *u* ∈ *H*. Then every *u* ∈ *H* is uniformly continuous. Furthermore, uniform equicontinuity implies equicontinuity.

EXAMPLES

1) Every finite set of continuous (resp. uniformly continuous) functions is equicontinuous (resp. uniformly equicontinuous); every finite union of equicontinuous (resp. uniformly equicontinuous) sets is equicontinuous (resp. uniformly equicontinuous).

2) If *E* and *F* are metric spaces, the set of all isometries from *E* into *F* is uniformly equicontinuous.

3) If *H* is a non-empty subset of $\mathscr{F}(E, F)$ we consider for every *x* ∈ *E* the mapping $u \mapsto u(x)$ from *H* into *F* that we write $\tilde{x}$; it is an element of $\mathscr{F}(H, F)$. If we put on $H \subset \mathscr{F}(E, F)$ the topology of the simple convergence or a finer topology, the mapping $u \mapsto u(x)$ is continuous (Section 4, Example 1), therefore $\tilde{x}$ belongs to $\mathscr{C}(H, F)$.

PROPOSITION 9

1) *Let E be a topological space and $H \subset \mathscr{F}(E, F)$. The subset H is equicontinuous at x_0 of E (or equicontinuous in E) if and only if the mapping $x \mapsto \tilde{x}$ from E into $\mathscr{C}_u(H, F)$ is continuous at x_0 (or continuous in E).*

2) *Let E be a uniform space and $H \subset \mathscr{F}(E, F)$. Then H is uniformly equicontinuous if and only if the mapping $x \mapsto \tilde{x}$ from E into $\mathscr{C}_u(H, F)$ is uniformly equicontinuous.*

3) *More generally, let f be a mapping continuous in each variable from a product $H \times E$ of two topological spaces into a uniform space F. If we denote by f_u, the mapping $x \mapsto f(u, x)$ from E into F and by $f_{.,x}$ the mapping $u \mapsto f(u, x)$ from H into F, then the following assertions are equivalent:*

a) *The set of $f_{u,.}(u \in H)$ is an equicontinuous (resp. uniformly continuous) subset of $\mathscr{C}(E, F)$.*

b) *The mapping $x \mapsto f_{.,x}$ from E into $\mathscr{C}_u(H, F)$ is continuous (resp. uniformly continuous).*

The proof follows directly from the definitions.

COROLLARY 1   *Let H be a set of mappings from a compact set E into a uniform space F. There is equivalence between H being equicontinuous or uniformly equicontinuous.*

This follows from the fact that a continuous mapping from a compact space E into a uniform space $\mathscr{C}_u(H, F)$ is *uniformly* continuous.

COROLLARY 2   *Let M be a topological (resp. uniform) space, $\mathfrak{S}$ a set of subsets of E. A mapping f from M into $\mathscr{F}_{\mathfrak{S}}(E, F)$ is continuous (resp. uniformly continuous) if and only if for every $A \in \mathfrak{S}$ the set of mappings*

$$t \longmapsto f(t)(x),$$

where $x \in A$ is an equicontinuous (resp. uniformly equicontinuous) set of mappings from M into F.

THEOREM 3   *Let E and F be topological spaces, G a uniform space and f a mapping from $E \times F$ into G.*

1) *If f is continuous, then*

   a) *for every $y \in F$, $f_{.,y}$ is continuous;*

   b) *the mapping $y \longmapsto f_{.,y}$ from F into $\mathscr{C}_c(E, G)$ is continuous.*
   b) *is equivalent to*

   b') *the set of $f_{x,.}$ where x runs through a compact set of E is an equicontinuous subset of $\mathscr{C}(F, G)$.*

2) *If E is locally compact, conditions a) and b) are also sufficient for the continuity of f.*

The equivalence of b) and b') is a particular case of Proposition 9, 3). If f is continuous so are the $f_{.,y}$. We prove b): Let $y \in F$, K be a compact set in E and U a symmetric entourage in G. For every $x \in K$ there exists an *open* neighborhood V_x of x and a neighborhood W_x of y such that for every $x' \in V_x$ and $y' \in W_x$ we have $(f(x', y'), f(x, y)) \in U$. Since K is compact there exists a finite number of points $x_i \in K$ ($1 \leqslant i \leqslant n$) such that the V_{x_i} form a covering of K. Indicating by W the intersection of neighborhoods W_{x_i} of y we have: for every $x' \in K$ there exists an $x_i \in K$ such that $x' \in V_{x_i}$, then $(f(x', y'), f(x_i, y)) \in U$ for every $y' \in W$ and in particular $(f(x', y), f(x_i, y)) \in U$. We conclude then that $(f(x', y'), f(x', y)) \in U$ for any $x' \in K$, which ends the proof of 1).

In order to show 2) we can reduce to the case where E is compact and consider f to be composed of the mapping $(x, u) \longmapsto u(x)$ from $E \times \mathscr{C}_u(E, G)$ into G, which is continuous (Proposition 8) and the mapping $(x, y) \mapsto (x, f_{., y})$ from $E \times F$ into $E \times \mathscr{C}_u(E, G)$ which is also continuous by hypothesis.

THEOREM 4 *Let E be a topological (resp. uniform) space and H an equicontinuous (resp. uniformly equicontinuous) subset of $\mathscr{C}(E, F)$. On H, the uniformity of simple convergence in E, of simple convergence in a dense subset E_0 of E and of compact (resp. precompact) convergence, are identical.*

It suffices to show that on H, simple convergence in E_0 and compact (resp. precompact) convergence are identical. Suppose E is a topological space; we must show that if U is an entourage in F and K a compact subset in E, there exists a finite subset A of E_0 and an entourage U' in F such that $W(A, U') \subset W(K, U)$. We choose for U' a symmetric entourage such that $U' \subset U$. Since H is equicontinuous there exists for every $x_0 \in K$ an open neighborhood V of x_0 in E such that for every $x \in V$ and $u \in H$, $(u(x), u(x_0)) \in U'$. Let $(V_i)_{1 < i < n}$ be a *finite* sequence of such neighborhoods covering K, we choose in each V_i an element $x_i \in E_0$ and consider the set A union of the x_i $(1 \leqslant i \leqslant n)$. Let $u, v \in H$ with $u, v \in W(A, U')$, i.e.

$$(u(x_i), v(x_i)) \in U'$$

for any $x_i \in A$, then we conclude that

$$(u(x), v(x)) \in U' \subset U$$

for any $x \in K$ which proves the first part. In the case where E is a uniform space and H uniformly continuous let K be a *precompact* subset of E, U an entourage in F and U' as above. There exists an entourage V in E such that for every $x, y \in K$, $(x, y) \in V$ and $u \in H$ we have

$$(u(x), u(y)) \in U'.$$

We now use Proposition 3. 1) and construct A as above.

THEOREM 5 *If H is an equicontinuous (resp. uniformly equicontinuous) subset of $\mathscr{F}(E, F)$ then its closure in $\mathscr{F}_s(E, F)$ is equicontinuous (resp. uniformly equicontinuous).*

The verification is immediate. Taking into consideration Theorem 4 we obtain:

COROLLARY *This closure is identical to the closure for the compact (resp. precompact) convergence.*

PROPOSITION 10 *Every precompact subset H of $\mathscr{C}_u(E, F)$ is equicontinuous; if furthermore E is a uniform space and if the elements of H are* *uniformly continuous functions, then H is uniformly equicontinuous. On H the uniform structures of simple convergence in a dense subset E_0 of E and of uniform convergence are identical.*

We must show that the mapping $x \mapsto \tilde{x}$ from E into $\mathscr{C}_u(H, F)$ is continuous (resp. uniformly continuous). Now the image $\tilde{E}$ of E in $\mathscr{C}_u(H, F)$ is uniformly equicontinuous (Proposition 8, Corollary 2), therefore, H being precompact, on $\tilde{E}$ the uniformly convergence uniform structure is identical to the simple convergence uniform structure (Theorem 4). It suffices then to verify that $x \mapsto \tilde{x}$ is continuous (resp. uniformly continuous) for simple convergence in $\mathscr{C}(H, F)$ which means precisely that the mappings $x \mapsto \tilde{x}(u) = u(x)$ are continuous (or uniformly continuous) on E. Finally, in the last assertion of Proposition 10 we can replace E_0 by E using Theorem 4, and it is then sufficient to apply Proposition 6, Corollary 2).

7 Relatively compact and precompact sets of continuous functions

THEOREM 6 (ASCOLI)   *Let E be a compact (resp. precompact) set, F a uniform space, H a set of continuous (resp. uniformly continuous) mappings from E into F.*

1) *H is precompact in $\mathscr{C}_u(E, F)$ if and only if H is equicontinuous (resp. uniformly equicontinuous) and H(x) is precompact for every $x \in E$.*

2) *Suppose F to be Hausdorff; H is relatively compact in $\mathscr{C}_u(E, F)$ if and only if H is equicontinuous (or uniformly equicontinuous) and H(x) is relatively compact for every $x \in E$.*

In the first statement the necessity follows from the preceding proposition and the uniform continuity of the $\tilde{x}$; the sufficiency follows from the fact that on H the uniform convergence and simple convergence uniformities are identical and from Proposition 3, 3). In the second statement we can suppose H closed. This is necessary as seen above. For the sufficiency we can suppose F complete (if not, we complete it and H will also be closed in $\mathscr{C}_u(E, \hat{F})$), whence $\mathscr{C}_u(E, F)$ is complete. Thus H will be precompact by 1), hence compact because it is complete.

COROLLARY 1   *Let E be a locally compact or metrisable space, F a Hausdorff uniform space and H a subset of $\mathscr{C}(E, F)$. Then H is relatively compact (resp. precompact) in $\mathscr{C}_c(E, F)$ if and only if H is equicontinuous and H(x) is relatively compact (resp. precompact) for every $x \in E$.*

These conditions are sufficient with no restriction on the space E: we consider for every compact K in E the set H_K of restrictions of all elements u of H to K and we use the remark following Proposition 4. To show the necessity we reconsider Theorem 6, and notice that a set H of mappings from E into F is equicontinuous if and only if for every compact $K \subset E$ the set H_K of restrictions to K of the $u \in H$ is equicontinuous.

COROLLARY 2 *Let E be a topological (resp. uniform) space, $\mathfrak{S}$ a set of subsets of E, F a uniform space and H a set of continuous (resp. uniformly continuous) mappings from E into F. For $A \in \mathfrak{S}$ let H_A be the set of restrictions to A of the $u \in H$. Then H is precompact in $\mathscr{C}_{\mathfrak{S}}(E, F)$ if and only if H_A is equicontinuous (resp. uniformly equicontinuous) for every $A \in \mathfrak{S}$ and $H(x)$ is precompact for every $x \in E_0 = \bigcup\limits_{A \in \mathfrak{S}} A$. If the $A \in \mathfrak{S}$ are precompact these conditions are also sufficient.*

The first part follows immediately from Proposition 9 and the second part from Theorem 6. 1), as it is sufficient to verify that the H_A are precompact for the uniform convergence.

REMARK We often wish to consider subspaces M of a space $\mathscr{C}_{\mathfrak{S}}(E, F)$ instead of $\mathscr{C}_{\mathfrak{S}}(E, F)$. The preceding chapter allows us to give criteria for relative compactness of subsets H of M. In fact, H is relatively compact *in M* if and only if it is relatively compact in $\mathscr{C}_{\mathfrak{S}}(E, F)$ and its closure in this space is contained in M.

CHAPTER 1

General properties

EXCEPT FOR Section 10 and part of Section 9, the particular structure of the fields of real or complex numbers plays no role in this chapter; these fields can be replaced by a valued field, provided certain obvious adaptations are made. For simplicity, however, we will suppose that the base field K is either the real field **R** or the complex field **C**. We will assume a good knowledge of linear algebra over a field K (see Bourbaki, *Algebra*, Chapter 2) and general topology (see Bourbaki, *Topologie Générale*).

We omit the proofs of assertions that follow easily from former results.

1 General definition of a topological vector space

DEFINITION 1 *A topological vector space (henceforward abbreviated to TVS) E is a vector space E endowed with a topology such that the mappings*

$$(x, y) \mapsto x + y \text{ and } (\lambda, x) \mapsto \lambda x$$

from $E \times E$ into E and from $K \times E$ into E are continuous ($E \times E$ and $K \times E$ having product topologies). Such a topology on a vector space is said to be compatible with the vector structure.

A TVS is in particular an abelian topological group, the related general results (incidentally, trivial) could be used in what follows. Let E be a TVS. Translations on E are homomorphisms and so are the non-zero homothetic transformations (which are also automorphisms of E). It follows from the former that if A and B are subsets of E where either is open, then $A + B$ is open, for it is the union of translations of an open set. If M is a topological space, and if f, g, λ, μ are continuous mappings from M into E, E, K and K respectively, then $f(t) + g(t)$, $\lambda(t)f(t)$ and more generally $\lambda(t)f(t) + \mu(t)g(t)$, are continuous mappings of M into E.

PROPOSITION 1 *For a TVS, E to be Hausdorff, it is necessary and sufficient that the set reduced to the origin $\{0\}$ be closed.*

The necessity is evident; for the sufficiency we observe that the diagonal of $E \times E$ is the pre-image of $\{0\}$ induced by the continuous mapping $(x, y) \mapsto x - y$, therefore E is closed if $\{0\}$ is closed.

The topology of a TVS is known once the neighborhood filter of the origin is known, as the neighborhood filter of any point x is obtained by the translation x. More precisely:

PROPOSITION 2 *For a filter $\mathscr{V}$ over a TVS E to be the neighborhood filter of the origin for a topology compatible with the vector structure, it is necessary and sufficient that it satisfies the following conditions:*

1) *$\mathscr{V}$ admits a base of balanced sets (V is balanced if $\lambda V \subset V$ for all scalars λ of norm $\leqslant 1$).*

2) *$\mathscr{V}$ is closed under non-zero homothetic transformations.*

3) *For all $U \in \mathscr{V}$, there exists $V \in \mathscr{V}$ such that $V + V \subset U$.*

4) *The $V \subset \mathscr{V}$ are absorbing (i.e. for every $x \subset E$, $\lambda x \in V$ for λ sufficiently small).*

2 Products, subspaces, quotients

Let E be a vector space, (E_i) a family of TVS, for all i, f_i a linear mapping from E into E_i. Then, the coarsest topology on E for which the f_i are continuous is compatible with the vector structure, that is, it makes E into a TVS. Thus, the pre-image of a linear mapping f from E into a TVSF, is a TVS topology on E; in particular, a vector subspace E of a TVS F is a TVS for the induced topology (E will be called a *topological vector subspace* of F). In a similar way, the product of a family (E_i) of TVS, having a vector product structure and a product topology, is a TVS called the *topological vector product* of the E_i.

The essential facts concerning quotient spaces are condensed in

THEOREM 1

1) *Let E be a TVS, F a vector subspace, ϕ the canonical mapping of E into E/F. Then the quotient topology of the topology of E makes E/F a TVS. ϕ is an open mapping of E into E/F. A fundamental system of neighborhoods of the origin in E/F is formed by the canonical images of the neighborhoods of the origin in E running through a fundamental system of neighborhoods. The space E/F is Hausdorff if and only if F is closed.*

2) *Let G be a vector subspace of E containing F; consider G/F as a vector subspace of E/F. Then the topology induced by E/F on G/F is identical to the quotient topology of G by F. The canonical isomorphism of $(E/F)/(G/F)$ with E/G defined as in algebra is also a TVS isomorphism.*

Proof

1) First, it is transparent that ϕ is an open mapping (that means that if $U \subset E$ is open, then $U + F$ is open, as pointed out in Section 1). Therefore, $(E/F) \times (E/F)$ is identified with $(E \times E)/(F \times F)$ and $K + (E/F)$ with $(K \times E)/(\{0\} \times F)$ (as topological spaces), which proves immediately the TVS axioms with respect to E/F. The characterization of the neighborhoods of the origin in E/F follows readily from the fact that ϕ is open; the criteria for E/F to be Hausdorff is a particular case of Proposition 1.

2) From the above, the sets $\phi(U \cap G)$, where U runs through the open sets of E are open sets of the quotient topology on G/F; the sets $\phi(U) \cap (G/F)$ are open sets of the topology induced by E/F on G/F. It follows immediately that $\phi(U \cap G) = \phi(U) \cap (G/F)$. The second part of 2) can be seen in a similar way or as a particular case of a result of general topology on quotient spaces.

Notice also that since the canonical mapping of a TVS onto a quotient space is open, *the classical algebraic isomorphism of $(E_1/F_1) \times (E_2/F_2)$ with $(E_1 \times E_2)/(F_1 \times F_2)$ is a* TVS *isomorphism* (E_1 and E_2 being TVS, F_1 and F_2 vector subspaces).

Care should be taken if ϕ is the canonical mapping of a TVS E onto a quotient space E/F, and G a vector subspace of E not containing F, as then the canonical mapping of $G/(G \cap F)$ onto the subspace $\phi(G)$ of E/F is not in general a TVS isomorphism (even in the case of an algebraic isomorphism and a continuous mapping). This means also that if ϕ is a homomorphism of E onto a space H (see Section 3), the mapping induced on a subspace G of E may well not be a homomorphism. This follows from the remark that every continuous linear mapping u from G into H can be induced by a homomorphism from $G \times H$ onto H, namely the mapping

$$(g, h) \longmapsto u(g) + h$$

3 Continuous linear mappings, homomorphisms

Let E, F be TVS, u a linear mapping from E into F. u is continuous if and only if it is continuous at the origin. In that case, if N is the kernel, u defines an injective continuous linear mapping from E/N into F. This mapping will not be in general an isomorphism from E/N onto $u(E)$ for the TVS structures, i.e. the inverse mapping is not always continuous.

DEFINITION 2   *A linear mapping u from a TVS E into a TVS F is a* *homomorphism, if the mapping from E/N (N, the kernel of u) onto $u(E)$ defined by u, is a TVS isomorphism.*

PROPOSITION 3 *Let u be a continuous linear mapping from a TVS E into a TVS F. The following conditions are equivalent:*

a) *u is a homomorphism.*

b) *u transforms the open subsets of E into open subsets of $u(E)$.*

c) *u transforms the neighborhoods of the origin of E into neighborhoods of the origin of $u(E)$.*

It follows immediately that a) implies b), b) implies c) and c) implies a).

The continuous linear mappings from E into F form a vector space, written $L(E, F)$.

4 Uniform structure of a TVS

Let E be a TVS. For every neighborhood U of the origin, let $\tilde{U}$ be the set of couples $(x, y) \in E \times E$ such that $x - y \in U$. When U varies, the $\tilde{U}$ form the basis of an entourage filter for a uniform structure on E compatible with the topology of E. Because of this uniform structure, E will implicitly be considered to be a uniform space.

The translations, the non-zero homothetics, are automorphisms of E considered as a uniform space.

If E and F are TVS, every linear mapping from E into F is already uniformly continuous. More generally, if a set A of linear mappings of E into F is equicontinuous at the origin, A is uniformly equicontinuous.

Let E be a TVS whose topology is the coarsest for which the linear mappings f_i from E into the TVS E_i are continuous (Section 1). Then the uniform structure of E is the coarsest for which the f_i are uniformly continuous; in particular, the uniform structure of a topological vector subspace of a TVS is the induced structure, and the uniform structure of the product of a family of TVS is the product of the uniform structures of the factor spaces. Thus, a subset of E is precompact if and only if the images by the f_i are precompact subsets of E_i.

The Hausdorff uniform space associated with a TVS E in the space E/N, where N is the closure of the origin.

The very definition of the uniform structure of a TVS E shows the following:

PROPOSITION 4 *The uniform structure of a TVS is metrisable if and only if E is Hausdorff and there exists a countable fundamental system of neighborhoods of the origin.*

(There will be, in fact, a countable fundamental entourage system.)

Since a TVS is also a uniform space, the meaning of a *complete* TVS is clear. For the spaces which are not complete, we have

PROPOSITION 5   *Let E be a Hausdorff TVS and $\hat{E}$ its uniform completion. Then the mapping $(x, y) \mapsto x + y$ from $E \times E$ into E can by continuity be extended to a continuous mapping from $\hat{E} \times \hat{E}$ into $\hat{E}$, and the mapping $(\lambda, x) \mapsto \lambda x$ from $K \times E$ into E can by continuity be extended to a continuous mapping from $K \times \hat{E}$ into $\hat{E}$. The laws thus defined in $\hat{E}$ make $\hat{E}$ a vector space and the topology of $\hat{E}$ is compatible with this vector structure. Finally, the uniform structure of the completion of E is merely the uniform structure associated with the TVS $\hat{E}$, which we have just defined.*

The proof is standard. We notice that $(x, y) \mapsto x + y$ is uniformly continuous in $E \times E$ and $(\lambda, x) \mapsto \lambda x$ uniformly continuous on the product of a bounded subset of K and E. From this we obtain the desired extension. The latter assertion is verified by calling F the TVS $\hat{E}$ and by using uniform continuity extending the identity isomorphism of E onto F to an isomorphism of the uniform space $\hat{E}$ onto the uniform space $\hat{F}$; this last mapping must be the identity mapping.

PROPOSITION 6   *Let E be a metrisable complete TVS and F a closed vector subspace. Then E/F is metrisable and complete.*

This is a particular case of Bourbaki, *Topologie Générale*, Chapter 9, Section 3, Proposition 4. We note that there exist complete locally convex spaces (not *metrisable*) not at all pathological, which admit non-complete quotient spaces. The validity of Proposition 6 is one reason for the importance of *metrisable* TVS.

5 Topology defined by a semi-norm

DEFINITION 3   *A semi-norm on a vector space E is a positive function p on E satisfying the conditions:*

$$p(x + y) \leqslant p(x) + p(y) \text{ for } x, y \in E$$
$$p(\lambda x) = |\lambda| \, p(x) \text{ for } x \in E, \lambda \in K.$$

The semi-norm p is a norm if additionally $p(x) = 0$ implies $x = 0$.

A semi-normed (resp. normed) space is a vector space which has a semi-norm (resp. a norm).

Very often, if E is a semi-normed space, its semi-norm is written $x \mapsto \|\,x\,\|$. The *unit ball* of E is the set of elements of E such that $\|\,x\,\| \leqslant 1$. If V is this set, the set of $x \in E$ such that $\|\,x\,\| \leqslant \lambda$ (where $\lambda > 0$) is λV. The set V is clearly balanced and absorbing from the second axiom of semi-norms, and we have $V + V \subset 2V$ from the first axiom. From Section 1 Proposition 1 we find:

PROPOSITION 6′ *Let E be a semi-normed space, V its unit ball. Then the non-zero homothetics λV of V are a basis for the neighborhood filter of a TVS topology on E. The function $\|\,x - y\,\|$ on $E \times E$ is a pseudo-metric which defines the uniform structure of the TVS E (which is therefore Hausdorff if and only if E is normed).*

A semi-normed space will therefore be considered as a TVS.

Subspaces, quotient spaces and product spaces of semi-normed spaces.

Let E be a semi-normed space, F a vector subspace. Then the semi-norm of E induces a semi-norm on F and the corresponding topology is the topology induced by E. Furthermore,

PROPOSITION 7 *Consider on E/F the function $X \mapsto \|\,X\,\|$ which to every $X \in E/F$ (i.e. a certain subset of E, translation of F) assigns the distance from the origin to X (relative to the pseudo-metric $\|\,x - y\,\|$)*

$$\|\,X\,\| = \inf_{x \in X} \|\,x\,\|$$

This function is a semi-norm on E/F, and the corresponding topology is the quotient topology of E by F.

From now on, a vector subspace or a quotient space of a semi-normed space will always be considered as a semi-normed space. In particular, the Hausdorff space E/N associated with a semi-normed space E (see Section 5) (where N is the closure of the origin, i.e. the set of $x \in E$ such that $\|\,x\,\| = 0$) can be considered to be a normed space, called the *normed space associated with the semi-normed space E*. This leads to the idea of a *metric homomorphism* of one semi-normed space into another: it is a continuous linear mapping of E into F such that the mapping of E/N onto the subspace $u(E)$ of F thereby obtained (where N is now the kernel of the mapping) is an isomorphism (for the structures of semi-normed spaces). A fortiori u will be a homomorphism in the sense of Section 3.

Let (p_i) be a family of semi-norms on a vector space. If $p(x) = \sup_i p_i(x)$

is finite for every $x \subset X$, then $p(x)$ is a semi-norm on E. The following is a more general way of constructing semi-norms from p_i. Consider the mapping

$$x \longmapsto \phi(x) = (p_i(x))$$

of E into R^I; let H be a vector subspace of R^I containing $\phi(E)$ and finally let α be a semi-norm on H which is *increasing* for the structure of natural order induced on H by R^I. Then

$$\alpha(p_i(x))$$

is a semi-norm on E. For instance, if the family (p_i) is finite, the functions

$$\sum_i p_i(x) \quad \text{and} \quad \left(\sum_i p_i(x)^2\right)^{1/2}$$

are semi-norms, since it is readily verifiable that the expressions

$$\sum_i |\xi_i| \quad \text{and} \quad \left(\sum_i \xi_i^2\right)^{1/2}$$

on R^n are norms. Notice finally that if F is a semi-normed space, and u a linear mapping from the vector space E into F, then $\| u(x) \|$ is a semi-norm on E, and the topology associated with it is the inverse image topology of E by u. These considerations lead to

PROPOSITION 8   *Let E be a vector space, (E_i) a finite family of semi-normed spaces and for every i, f_i a linear mapping of E into E_i. Then the coarsest topology on E for which the f_i are continuous may be defined by a semi-norm. We can use, for instance, any of the following semi-norms (where $p_i(x) = \| f_i(x) \|$):*

$$\sup_i p_i(x), \ \sum_i p_i(x), \ \left(\sum_i (p_i(x))^2\right)^{1/2}.$$

But a priori there should be no preferences, the choice being determined by convenience in each particular case. We see that in particular the topology of the product of a finite number of semi-normed (resp. normed) spaces may be defined by a semi-norm (resp. a norm).

The completion of a normed space   A complete normed space is called a *Banach space*. Let E be a normed space, $\| x \|$ (distance to the origin) is known to be uniformly continuous and by the continuity extendable to the completed vector space $\hat{E}$ (Section 4, Proposition 5) to a function, also written $\| x \|$, which is still a semi-norm. On the other hand, the function $\| x - y \|$ on $\hat{E} \times \hat{E}$, which by continuity extends the function $\| x - y \|$ on $E \times E$, is a distance on the completion of the metric space E, the topology of $\hat{E}$ is therefore the topology defined by its norm. Thus, *the completion of a normed space may be considered as a normed space.*

Continuous linear mappings between semi-normed spaces

THEOREM 2 *Let E, F be semi-normed spaces.*

1) *Let u be a linear mapping of E into F. Then u is continuous if and only if there exists an $M \geqslant 0$ such that*

$$|| u(x) || \leqslant M || x || \text{ for every } x \in E,$$

which is equivalent to

$$\sup_{||x|| \leqslant 1} || u(x) || \leqslant M$$

or to $$u(V) \subset MW$$

where V and W are the unit balls of E and F.

2) *For every $u \in L(E, F)$ let*

$$|| u || = \sup_{||x|| \leqslant 1} || u(x) ||.$$

The function thus defined on $L(E, F)$ is a semi-norm, and a norm if F is Hausdorff. With this norm, $L(E, F)$ is complete if F is complete.

If E and F are semi-normed spaces, $L(E, F)$ is always considered as a semi-normed space as indicated above. In any case, $|| u ||$ will be called (by abuse of the term) the norm of u.

Proof

1) It is evident that the first formula implies the others and that the last ones are equivalent. On the other hand

$$u(V) \subset MW$$

implies $u(\varepsilon V) \subset \varepsilon MW$ for every $\varepsilon > 0$, thus u is continuous at the origin and therefore continuous (Section 3). Conversely, if u is continuous, we will have $u(\varepsilon V) \subset W$ for ε sufficiently small and then $u(V) \subset MW$ with $M = 1/\varepsilon$. Finally, it is easy to see, by considering homogeneity, that the second formula in 1) implies the first.

2) For every $x \in E$, $u \mapsto || u(x) ||$ is a semi-norm on $L(E, F)$, since it is the inverse image of the semi-norm on F by the linear mapping $u \mapsto u(x)$. On the other hand, we have stated that every least upper bound of semi-norms is a semi-norm and therefore $|| u ||$ is a semi-norm on $L(E, F)$. The rest of the proof is left to the reader.

COROLLARY   *Two semi-norms p and q on the same vector space E define the same topology if and only if there exist two numbers m and M such that*

$$0 < m \leqslant M < +\infty$$
$$mp(x) \leqslant q(x) \leqslant Mp(x) \text{ for } x \in E.$$

We must express that the identity mapping of E with p onto E with q, as well as the inverse mapping, is continuous, and we apply the criterion of the preceding theorem.

6 Generalities concerning spaces defined by families of semi-norms

Let E be a vector space, (p_i) a family of semi-norms on E. Consider on E the least upper bound of the topologies associated with the p_i, which makes E a TVS (Section 2). The uniform structure of E is the least upper bound of the uniform structure associated with the p_i (Section 4), and therefore is defined by the families of pseudo-metrics $p_i(x - y)$. A fundamental family of neighborhoods of the origin in E is obtained by taking for each *finite* subset J of the set of indices I and for each $\varepsilon > 0$, the set of those x such that $p_J(x) \leqslant \varepsilon$ where we have set

$$p_J(x) = \sup_{i \in J} p_i(x).$$

In consequence, the topology defined by the family $(p_i)_{i \in I}$ of semi-norms remains unchanged when the least upper bound p_J of a finite number of such semi-norms is added. We can add the semi-norms on E which are bounded by a multiple of such a p_J; we verify that we have then obtained all the semi-norms on E which are continuous (the addition of other semi-norms would make the topology of E strictly finer).

DEFINITION 4   *A locally convex space is a TVS whose topology can be defined by a family of semi-norms (p_i) on E as described above (i.e. it is the least upper bound of the semi-normed topologies associated with the p_i). The family (p_i) is said to be a family of definition for the topology of E. Such a family is said to be fundamental if the unit balls associated with the p_i form a fundamental system of neighborhoods of the origin.*

   (This means also that for each finite subset J of I and for each $M > 0$, there exists a $p_i \geqslant Mp_J$.)

Nearly all TVS used in analysis are locally convex since they are the only spaces for which we have a sufficient number of important results. They are largely sufficient, since they are a family of spaces closed under the usual operations, as we will now see.

Let E be a TVS whose topology is defined as the coarsest for which the f_i from E into locally convex spaces E_i are continuous. Then E is locally convex; more precisely, its topology is defined by the family of semi-norms pf_i, where for all i, p runs through a family of definition for the topology of E_i (this follows from an obvious transitivity property, valid in general topology, for the method of definition of a topology as the coarsest which . . .).

Thus, a *topological vector subspace of a locally convex space, and the topological vector product of a family of locally convex spaces, are locally convex.* Let E be a locally convex space with a fundamental family of semi-norms, let F be a vector subspace and consider on E/F the *quotient* semi-norms $\dot{p}_i$ of the p_i (Section 5, Proposition 7). From Proposition 7, it follows that these semi-norms define precisely the quotient topology of E/F; therefore, *a quotient space of a locally convex space is locally convex.* In particular, the Hausdorff space associated with a locally convex space (Section 4) is locally convex. *The completion of a locally convex space is locally convex.* For, if (p_i) is a family of semi-norms on E defining the topology of E, the (p_i) can, by uniform continuity, be extended into semi-norms $\hat{p}_i$ on the completion, and the corresponding pseudo-metrics on $\hat{E}$, extensions by continuity of the pseudo-metrics $p_i(x - y)$ on E, define the uniform structure of $\hat{E}$ (*Topologie Générale*); the topology of $\hat{E}$ is therefore defined by the $\hat{p}_i$.

The following is a useful characterization of locally convex spaces:

PROPOSITION 9 *The topology of a locally convex space E can be defined as the coarsest for which certain linear mappings f_i from E into Banach spaces E_i are continuous. If in particular E is Hausdorff, E is isomorphic to a topological vector subspace of a topological vector product of Banach spaces.*

For, if (p_i) is a family of definition of the topology of E, it suffices to call E_i the Banach space completion of the normed space associated with the semi-norm p_i (Section 5), and f_i the canonical mapping of E into E_i.

7 Bounded sets: General criteria

DEFINITION 5 *Let E be a TVS. A subset A of E is bounded if for every neighborhood V of the origin there exists a scalar $\lambda \neq 0$ such that $\lambda A \subset V$.*

If A is bounded, we will have also $\lambda A \subset V$ provided λ is small enough (choose V balanced). A subset of a bounded set is bounded. Every subset of E reduced to a point is bounded (since the neighborhoods of the origin are absorbing); therefore, every finite subset of E is bounded as

c

the union, and the sum of a finite number of bounded sets are bounded (using Proposition 2. 3)). In particular, the translations of a bounded set are bounded; clearly the homothetics of a bounded set are also bounded. *The closure of a bounded set is bounded* (use the fact that E, being uniformizable, admits a fundamental system of closed neighborhoods of the origin).

PROPOSITION 10　*A precompact subset A of a* TVS *is bounded.*

For, if V is a balanced neighborhood of the origin, there exists a finite subset B of E such that $A \subset B + V$ (this is a statement of the precompactness of A). There exists also a scalar λ of norm $\leqslant 1$ such that $\lambda B \subset V$, from which

$$\lambda A \subset \lambda B + \lambda V \subset V + V.$$

Since V is arbitrary, A is bounded.

PROPOSITION 11　*Let E be a* TVS *whose topology is the coarsest for which certain linear mappings f_i from E into* TVS E_i *are continuous. Then a set A is bounded in E if and only if for every i, $f_i(A)$ is bounded in E_i.*

The necessity is a particular case of the (immediate) fact: *the image of a bounded set by a continuous linear mapping is bounded*. For the sufficiency we use the characterization of the neighborhoods of the origin in E. Particular cases to be made explicit are subspace and product space.

Let E be a semi-normed space, V its unit ball. $A \subset E$ is bounded if there exists $\lambda \neq 0$ such that $\lambda A \subset V$; the condition is also sufficient since we have $\varepsilon \lambda A \subset \varepsilon V$ for every $\varepsilon > 0$. Since $\lambda A \subset V$ can be written $A \subset MV$ where $M = 1/\lambda$ or $\| x \| \leqslant M$ for every $x \in A$, we see that:

PROPOSITION 12　*If E is a semi-normed space, the bounded subsets of E are the bounded subsets in the sense of the metric (more exactly in the sense of the pseudo-metric $\| x - y \|$), i.e., the subsets A such that*

$$\sup_{x \in A} \| x \| < \infty.$$

In particular, the unit ball of a semi-normed space is a bounded neighborhood of the origin. Conversely:

COROLLARY 1　*The topology of a* TVS E *may be defined by a semi-norm, or a norm respectively (we say that E is semi-normable, or normable), if and only if E is locally convex (resp. locally convex and Hausdorff), and admits a bounded neighborhood of the origin.*

For if V is a bounded neighborhood of the origin, and if p is a continuous semi-norm on E such that the corresponding unit ball U be contained in V, the topology of E is that defined by p. If W is any given neighborhood of the origin, we will have $\lambda V \subset W$ for some $\lambda \neq 0$, and a fortiori $\lambda U \subset W$.

COROLLARY 2 *Let u be a linear mapping of a semi-normed space E into a TVS F. Then u is continuous if and only if u transforms the unit ball of E into a bounded subset of F.*

The necessity follows from the fact that V is bounded. Conversely, if $u(V)$ is bounded and W a neighborhood of the origin in F, we will have

$$\lambda u(V) \subset W$$

for some $\lambda \neq 0$, from which $u(\lambda V) \subset W$. Since λV is a neighborhood of the origin in E, it follows that u is continuous at the origin, therefore continuous.

Now let E be a *locally convex space*, whose topology is defined by a family (p_i) of semi-norms. Combining Propositions 11 and 12 we see that a subset A is bounded if and only if for every i, we have

$$\sup_{x \in A} p_i(x) < +\infty.$$

We therefore easily conclude:

PROPOSITION 13 *Let E be a locally convex space. Then the disked hull of a bounded subset of E is bounded. (The disked hull of A is the set of finite sums $\Sigma \lambda_i x_i$, where the $x_i \in A$ and $\Sigma |\lambda_i| \leqslant 1$).*

Consequently, the *closed disked hull* of a bounded set (i.e. the closure of the disked hull) is a bounded set.

DEFINITION *A TVS is quasi-complete if its closed and bounded subsets are complete.*

EXERCISE Let E be a TVS. Show the following: a line of E is bounded if and only if it is contained in the closure of the origin. Therefore E is Hausdorff if and only if E does not contain bounded lines.

8 Bounded sets: Their use in $\mathfrak{S}$-convergences

The need for bounded sets appears in

THEOREM 3 *Let M be a set, $\mathfrak{S}$ a set of subsets of M, E a TVS, H a vector subspace of the space of all mappings of M into E with the $\mathfrak{S}$-convergence topology. This topology is compatible with the vector structure if and only if every $u \in H$ transforms the $A \in \mathfrak{S}$ into bounded subsets of E. Then the* *uniform structure of the* TVS H *is identical to the* $\mathfrak{S}$-*convergent uniform structure. If the topology of E is defined by a family $(p_i)_{i \in I}$ of semi-norms, the topology of H is defined by the semi-norms*

$$p_{i,A}(u) = \sup_{t \in A} p_i(u(t)),$$

where i runs through I and A through $\mathfrak{S}$ (therefore H is locally convex if E is). In particular, if $\mathfrak{S}$ is reduced to the unique element M (in the case of uniform convergence), and if E is semi-normed (or normed), then the topology of H is defined by the semi-norm (or the norm)

$$\| u \| = \sup_{t \in M} \| u(t) \|$$

(*called the uniform norm*).

The proof follows immediately by using for example Proposition 1 (Section 1). The four conditions on the neighborhood filter of the origin in H are clearly always satisfied, except for the last one, which demands that the $u(A)$ ($u \in H$, $A \in \mathfrak{S}$) be bounded subsets of E.

Theorem 3 leads to the consideration of the space of all mappings u of M into E which transform the $A \in \mathfrak{S}$ into bounded sets. With the $\mathfrak{S}$-convergent topology, it is a TVS denoted by $B_{\mathfrak{S}}(M, E)$. The TVS such as H are therefore the topological vector subspaces of $B_{\mathfrak{S}}(M, E)$. If in particular M is a topological space, we call $C_{\mathfrak{S}}(M, E)$ the topological vector subspace of $B_{\mathfrak{S}}(M, E)$ formed by the continuous mappings (bounded on the $A \in \mathfrak{S}$). When in particular $\mathfrak{S}$ is the set of compact subsets of M, we obtain the space of *all* continuous mappings of M into E (since the precompact subsets of E are bounded) (Proposition 10), with the compact convergence topology: this space is written $C(M, E)$. When $\mathfrak{S}$ is reduced to the single element M (in the case of uniform convergence), we denote the corresponding space by $C^{\infty}(M, E)$, i.e. the space of continuous and bounded mappings of M into E, with the topology of uniform convergence. If E is a normed space, $C^{\infty}(M, E)$ is always implicitly considered as a space normed by the uniform norm

$$\| u \| = \sup_{t \in M} \| u(t) \|.$$

When M is compact, the spaces $C^{\infty}(M, E)$ and $C(M, E)$ coincide. Finally, when E is the field of scalars, we omit it from the expression and write $C_{\mathfrak{S}}(M)$, $C(M)$, $C^{\infty}(M)$ (the latter is always a *Banach space*: see the following proposition).

PROPOSITION 14 *Using the preceding notation, if E is complete, $B_{\mathfrak{S}}(M, E)$ and $C^{\infty}(M, E)$ are complete. This is equally true for $C(M, E)$ if for example M is locally compact or metrisable.*

We know that if E is complete, the space of all mappings of M into E with a $\mathfrak{S}$-convergence uniformity is complete; it suffices to show that the spaces mentioned in the proposition are *closed* subspaces of the above-mentioned space. This is well known for $C(M, E)$ when M is metrisable or locally compact. Moreover this can be directly verified for $B_{\mathfrak{S}}(M, E)$. Furthermore, $C^{\infty}(M, E)$, the intersection of the space $B_{\mathfrak{S}}(M, E)$ (where $\mathfrak{S}$ is reduced to M) and the space (also closed) of all continuous functions with values in E, is also closed.

If M and E are semi-normed spaces, and if we consider on $L(M, E)$ the topology of uniform convergence on the unit ball V of M, this topology is defined by the norm

$$\sup_{\|x\| \leqslant 1} \| u(x) \|,$$

which is simply the norm defined in Section 5, Theorem 2. Since every limit for simple convergence (and a fortiori for uniform convergence on V) of linear mappings of M into E is clearly linear (at least if E is Hausdorff), Proposition 14 implies that $L(M, E)$ is complete if E is a Banach space.

PROPOSITION 15 *Under the conditions of Theorem 3, a subset P of H is bounded if and only if for every $A \in \mathfrak{S}$, the set $P(A) = \bigcup\limits_{u \in P} u(A)$ is a bounded subset of E.*

EXERCISE Show that if M is locally compact but not compact, the topology of $C(M)$ cannot be defined by a semi-norm (use Proposition 12, Corollary 2).

9 Examples of TVS: Spaces of continuous functions

An isomorphism and a canonical homomorphism Let M be a normal space, N a *closed* subspace. The mapping which to every $f \in C^{\infty}(M)$ assigns the restriction of f to N is evidently a linear mapping, of norm $\leqslant 1$, of $C^{\infty}(M)$ into $C^{\infty}(N)$. Uryson's theorem implies that *we even have a metric homomorphism from $C^{\infty}(M)$ onto $C^{\infty}(N)$*, therefore identifying the latter space (as a normed space) with the quotient of $C^{\infty}(M)$ by the subspace J of functions that are zero on N.

Let K and L be locally compact spaces, E a TVS. Then the spaces $C(K \times L, E)$ and $C(K, C(L, E))$ are canonically isomorphic as uniform spaces (this is true whenever E is a uniform space), and this isomorphism respects the vector structure. *Thus, $C(K \times L, E)$ is identified with $C(K, C(L, E))$ as a TVS.* This isomorphism respects the norms when E

is normed. In particular, we have a canonical isomorphism of normed spaces:

$$C(K \times L) \approx C(K, C(L)).$$

We draw attention to the fact that in general, if L is not compact, we will have $C^\infty(K \times L) \neq C^\infty(K, C^\infty(L))$; a priori, the second space is contained in the first but in general strictly contained (the reader will find examples).

An approximation property

THEOREM 4   *Let M be a compact space, E a locally convex space. Then the linear combinations of functions $\phi . a$ (where $\phi \in C(M)$, $a \in E$) are dense in $C(M, E)$.*

Proof   Let $f \in C(M, E)$. We must show that for every continuous semi-norm p on E, there exists a function $g = \sum \phi_i a_i$ such that $p(f(t) - g(t)) \leqslant 1$ for every $t \in M$. Since f is uniformly continuous, there exists a finite covering (U_i) of M by open sets and a point t_i contained in each U_i such that $t \in U_i$ implies

$$p(f(t) - f(t_i)) \leqslant 1.$$

Set $f(t_i) = a_i$, let (ϕ_i) be a partition of unity subordinated to (U_i) and choose

$$g = \sum_i \phi_i a_i.$$

We will then have $f(t) - g(t) = \sum_i \phi_i(t)(f(t) - f(t_i))$; where we can restrict the sum to those i such that $t \in U_i$ and obtain

$$p(f(t) - g(t)) \leqslant 1.$$

COROLLARY   *Let K and L be compact spaces. Then the functions which are linear combinations of functions of type $f(s)g(t)$ (where $f \in C(K)$, $g \in C(L)$), are dense in $C(K \times L)$.*

It is sufficient to see that

$$C(K \times L) = C(K, C(L)).$$

From Theorem 4 we see that this theorem is valid when M is only locally compact.

Separability properties

PROPOSITION 16   *Let M be a compact space. M is metrisable if and only if $C(M)$ is separable.*

Sufficiency Let (f_i) be dense in $C(M)$. If s, t are two points of M such that $f_i(s) = f_i(t)$ for every i, by continuity we will have $f(s) = f(t)$ for every $f \in C(M)$, therefore $s = t$. The coarsest topology on M for which the f_i are continuous is therefore Hausdorff and consequently identical to the topology of M since it is coarser, but this topology is metrisable.

Necessity Consider a sequence of finite open coverings of M by sets whose diameter tends to zero and for each of these coverings a subordinate partition of unity. The proof of Theorem 4 shows that the linear combinations of functions ϕ_i corresponding to the unit partitions are dense in $C(M)$. This is also true for rational linear combinations of these functions, and the conclusion follows.

Combining Proposition 16 with Theorem 4 we obtain the

COROLLARY *Let K be a compact metrisable space, E a separable locally convex space. Then $C(K, E)$ is separable.*

The space $C_0(M)$.

Let M be a locally compact space, E a TVS. A mapping f of M into E is said to be *zero at infinity* if it tends to zero following the filter of the complements of compact subsets of M. We wrote $C_0(M, E)$ for the space of continuous mappings of M into E zero at infinity and having the uniform convergence topology. This is clearly a TV subspace of $C^\infty(M, E)$. Let $\hat{M}$ be the one point compactification of M (supposed not compact), let ω be the point at infinity of $\hat{M}$ (i.e. the only point of $\hat{M}$ which does not belong to M). *A continuous function from M into E is zero at infinity if and only if the extension to $\hat{M}$ obtained when ω has the value zero, is continuous.* From this, $C_0(M, E)$ is identified with the closed subspace of $C(\hat{M}, E)$ formed by functions which are zero at ω. This identification respects the topologies. The constant mappings of M into E form a natural supplement of $C_0(M, E)$.

PROPOSITION 17 *Let M be a locally compact space, E a locally convex space. Then $C_0(M, E)$ is the closure in $C^\infty(M, E)$ of the space of continuous mappings having a compact support of M in E (the support of a function having vectorial values is the closed set complementary to the set of points in whose neighborhood the function is zero).*

EXERCISE 1 Show that under the preceding conditions (M a normal space, N a closed subspace), if E is a Banach space, the canonical mapping of $C^\infty(M, E)$ into $C^\infty(N, E)$ is a metric homomorphism of the first space onto a closed subspace of the second (use Uryson's theorem and Proposition 6, Section 4). Conclude from this that it is a metric homomorphism of $C^\infty(M, E)$ onto $C^\infty(N, E)$ if N is compact (use

Theorem 4 above). Prove the analogous proposition when E is a locally convex metrisable and complete space. What can we say when E is any locally convex space?

EXERCISE 2   Let M be a completely regular space. E a locally convex space. Show that the functions $f \in C^\infty(M, E)$ belonging to the closed vector space generated by the functions $\phi . a$ ($\phi \in C^\infty(M)$, $a \in E$), are exactly those for which $f(M)$ is precompact. (Proceed as in Theorem 4 or use Theorem 4 introducing the *Stone compactification* of M. See Bourbaki, *Topologie Générale*, Chapter 9, Section 1, Exercise 6).

EXERCISE 3   Let M be a compact space, E a metrisable and separable uniform space. Show that $C(M, E)$ is a metrisable and separable space. (Choose a metric for E and prove first of all the lemma: a separable metric space E is always isomorphic to a subspace of a separable Banach space. To see this, take a dense sequence (x_i) in E, $a \in E$ arbitrary and show that the mapping $x \mapsto \phi(x) = (d(x, x_i) - d(a, x_i))_{i \in I}$ into the space $C^\infty(N)$ of bounded sequences is an isometry; then take the closed vector space generated by $\phi(E)$.)

EXERCISE 4   If M is a completely regular non-compact space, then $C^\infty(M)$ is not separable. (Otherwise M would be a dense subspace of a compact metrisable space $\hat{M}$ (its stone compactification), different from M, and such that every bounded continuous function on M could be extended by continuity to all of $\hat{M}$.)

EXERCISE 5   Apply the preceding identification in order to characterize the relatively compact subsets of $C_0(M, E)$. Case $C_0(M)$. Extend Theorem 4 to the spaces $C_0(M, E)$.

EXERCISE 6   Let I be an index set, let $l^1(I)$ be the space of summable families of scalars over the index set I:

a) Show that the function

$$\| (\lambda_i) \|_1 = \Sigma \, | \, \lambda_i \, |$$

over $l^1(I)$ is a norm, which makes it a Banach space.

b) Let E be a complete TVS. Show that if (x_i) is a bounded family in E, then for every $\lambda = (\lambda_i) \in l^1(I)$, the family $(\lambda_i x_i)$ is a summable family in E, and its sum $u(\lambda)$ is a continuous linear function of $\lambda = (\lambda_i) \in l^1(I)$. Show that we thus obtain exactly all the continuous linear mappings of $l^1(I)$ into E.

c) If E is a Banach space, the norm of the linear mapping of $l^1(I)$ into E defined by means of the sequence (x_i) is identical to $\sup_{i \in I} \| x_i \|$.

EXERCISE 7 Let I be an index set. We denote by $l^\infty(I)$ and $c_0(I)$ the spaces $C^\infty(I)$ and $C_0(I)$ built upon I considered as a discrete locally compact space; they are therefore Banach spaces and $c_0(I)$ is a normed subspace of $l^\infty(I)$. On the other hand, $l^1(I)$ represents the space of scalar functions $i \mapsto \lambda_i$ on I such that $\Sigma \,|\, \lambda_i \,| < +\infty$ with the norm $\Sigma \,|\, \lambda_i \,|$ (see the preceding exercise). When I is implicit or is the set of positive integers, we write the preceding spaces simply l^∞, c_0 and l^1. If $(\lambda_i) \in l^\infty$ and $(\mu_i) \in l^1$, set

$$\langle (\lambda_i), (\mu_i) \rangle = \Sigma \, \lambda_i \mu_i.$$

Show that the dual (i.e. the space of continuous linear mappings having a natural norm—see Section 5) of c_0 can be identified with l^1, and the dual of l^1 can be identified with l^∞ (for the dual of l^1 see the preceding exercise).

10 Other examples: The spaces $\mathscr{E}^{(m)}$ and $\mathscr{E}$ of L. Schwartz

See *Théorie des Distributions* by L. Schwartz for the notation D^p for the derivation operators on R^n, where $p = (p_1, \ldots, p_n)$ is any system of n integers $\geqslant 0$. D^p stands for the operator

$$\frac{\partial^{|\,p\,|}}{\partial x_1^{p_1} \ldots \partial x_n^{p_n}}$$

where $|\, p \,| = p_1 + \ldots + p_n$ is the order of this differential operator. We have $D^0 = \text{identity}$, $D^p D^q = D^{p+q}$.

Let U be an open non-empty subset of R^n, call

$$\mathscr{E}^{(m)}(U)$$

or simply $\mathscr{E}^{(m)}$ the space of scalar functions on U which are m times continuously differentiable, having the coarsest topology for which the D^p ($|\, p \,| \leqslant m$) from this space into $C(U)$ are continuous ($C(U)$ is the space of continuous functions on U with a compactly convergent topology). We thus obtain a *locally convex* and *metrisable* space (Section 4, Proposition 4), where a fundamental sequence of semi-norms is obtained by taking a "fundamental sequence" (K_n) of compact subsets of U and by setting

$$p_n(f) = \sup_{|p| \leqslant m, \, t \in K_n} |\, D^p f(t) \,|.$$

A filter (f_i) in $\mathscr{E}^{(m)}$ converges towards $f \in \mathscr{E}^{(m)}$ if and only if for every p with $|\, p \,| \leqslant m$, the $D^p f_i$ converge towards $D^p f$ uniformly on every compact set. Let (f_i) be a Cauchy filter in $\mathscr{E}^{(m)}$, then for every p of order $\leqslant m$, the $D^p f_i$ form a Cauchy filter in $C(U)$, thus converge in this space towards a $g_p \in C(U)$ (Section 8, Proposition 14). We can show that g_0 is m times continuously differentiable, that $D^p g_0 = g_p$ and that if every partial derivative $(\partial/\partial x)f_i$ tends to a limit h uniformly on every compact set, g_0 is continuously differentiable and its derivatives are the h_a. To see this we can consider the case of one variable where the result is classical.

We have thus proved that $\mathscr{E}^{(m)}$ is *complete*.

We call $\mathscr{E}(U)$ or simply $\mathscr{E}$ the intersection of the spaces $\mathscr{E}^{(m)}(U)$ having a least upper bound topology with respect to the induced topologies. This is also the coarsest topology for which the D^p from $\mathscr{E}(U)$ into $C(U)$ are continuous (where p now runs through the set of all indexes of derivation). It is furthermore a locally convex and metrisable space and the above reasoning shows that it is complete. This can be shown also from the following useful lemma:

LEMMA   *Let F be a set, (E_m) a decreasingly directed family of subsets with uniform structures such that for $E_m \subset E_n$, the uniform structure of E_m is finer than that induced by E_n. Let E be the intersection of the E_m with the least upper bound uniform structure of the structures induced by the E_m. If the E_m are Hausdorff and complete, so is E.*

Suppose now that K is a compact cube of R^n. We can define as above the spaces $\mathscr{E}^{(m)}(K)$, $\mathscr{E}(K)$, and the preceding reasoning shows as before that they are locally convex metrisable and complete spaces. The essential difference is that the $\mathscr{E}^{(m)}(K)$ are normable since $C(K)$ is a Banach space (Section 5, Proposition 8), while it is easy to see that the $\mathscr{E}^{(m)}(U)$ are not normable (use Proposition 12, Corollary 1); $\mathscr{E}(U)$ and $\mathscr{E}(K)$ are not normable.

In Section 7, Proposition 11 we have given a criterion for the bounded subsets of $\mathscr{E}^{(m)}(U)$ or $\mathscr{E}(U)$; a subset A of $\mathscr{E}^{(m)}(U)$ (or of $\mathscr{E}(U)$) is bounded if and only if for every index of derivation p of order $\leqslant m$ (or of any order) the set $D^p(A)$ is a bounded subset of $C(U)$ (i.e. Section 8, Proposition 15—a set of continuous functions on U, *uniformly bounded on every compact set in U*). Also, a subset A of $\mathscr{E}^{(m)}(U)$ is relatively compact if and only if it is precompact, therefore if for every p of order $\leqslant m$, the set $D^p(A)$ is a precompact subset of $C(U)$. Applying Ascoli's theorem we find

PROPOSITION 18   *Let A be a subset of $\mathscr{E}^{(m)}(U)$. A is relatively compact if and only if A is bounded and for every index of derivation p of order m the set of functions $D^p(A)$ is equicontinuous.*

It suffices to show that for every p of order $\leqslant m$ the set of functions $D^p(A)$ is equicontinuous. This follows from:

LEMMA *Let B be a set of continuously differentiable functions on U uniformly bounded on every compact set together with their first order derivatives. Then B is equicontinuous. (Mean Value Theorem).*

Combining Proposition 18 and the lemma we get

COROLLARY *Every bounded subset of $\mathscr{E}^{(m+1)}(U)$ is relatively compact in $\mathscr{E}^{(m)}(U)$.*

A fortiori, every bounded subset of $\mathscr{E}(U)$ is relatively compact in all of the $\mathscr{E}^{(m)}(U)$, from this we obtain the important

THEOREM 5 *In $\mathscr{E}(U)$, bounded subsets and relatively compact subsets are identical.*

All of the preceding can be repeated word for word for the spaces

$$\mathscr{E}^{(m)}(K) \quad \text{and} \quad \mathscr{E}(K),$$

in particular every bounded subset of $\mathscr{E}(K)$ is relatively compact. Since $\mathscr{E}^{(m+1)}(K)$ is normable, the Corollary of Proposition 18 shows that the identity mapping of $\mathscr{E}^{(m+1)}(K)$ into $\mathscr{E}^{(m)}(K)$ transforms some neighborhood of the origin into a relatively compact subset of $\mathscr{E}^{(m)}(K)$. We say that it is a *compact mapping*.

EXERCISE 1 Define the spaces $\mathscr{E}^{(m)}(U, E)$, $\mathscr{E}(U, E)$, $\mathscr{E}^{(m)}(K, E)$, $\mathscr{E}(K, E)$ when E is any TVS. Show that these spaces are complete when E is locally convex and complete (use Section 6, Proposition 9 for the case E a Banach space). Show that if V is open in R^n, L a compact cube of R^n, we have a TVS isomorphism

$$\mathscr{E}^{(m)}(U \times V) = \mathscr{E}^{(m)}(U, \mathscr{E}^{(m)}(V))$$
$$\mathscr{E}^{(m)}(K \times L) = \mathscr{E}^{(m)}(K, \mathscr{E}^{(m)}(L))$$

(similarly for $\mathscr{E}(U \times V)$ and $\mathscr{E}(K \times L)$).

EXERCISE 2 Characterize the relatively compact subsets of $\mathscr{E}^{(m)}(U, E)$ and of $\mathscr{E}(U, E)$, when E is locally convex, Hausdorff and complete. Under what conditions is every bounded subset of $\mathscr{E}(U, E)$ relatively compact?

EXERCISE 3 If $m > m'$, the identity mapping of $\mathscr{E}^{(m)}(U)$ into $\mathscr{E}^{(m')}(U)$, is not a TVS isomorphism.

11 Topological direct sums

Recall that two subspaces F, G of a vector space E are *supplementary* if the mapping $(x, y) \mapsto x + y$ of $F \times G$ into E is an isomorphism, which means in particular that $F \cap G = \{0\}$, $F + G = E$ (Direct sum). The projection of $F \times G$ onto F defines a linear mapping p of E onto F which considered as an endomorphism of E satisfies

$$p^2 = p, \quad p(E) = F, \quad p^{-1}(0) = G.$$

The analogous mapping of E onto G is $1 - p$. Conversely, if p is an endomorphism of E such that $p^2 = p$, we see that p can be obtained as described above and this uniquely, since we will have $F = p(E)$, $G = p^{-1}(0)$. An endomorphism p of a vector space E such that $p^2 = p$ is called a *projector*. The projectors of E therefore correspond exactly to the decompositions of E in direct sums of 2 supplementary subspaces F, G.

Suppose now that E is a TVS. If F and G are supplementary subspaces of E, the algebraic isomorphism of $F \times G$ onto E is clearly continuous but not always a TVS isomorphism, i.e. the inverse mapping is not always continuous. If we have a TVS isomorphism we say that they are *topological supplements*.

PROPOSITION 19 *Let E be a TVS, F a vector subspace.*

1) *A projector p of E onto F defines a topological supplement G if and only if p is continuous. Then $1 - p$ is a topological homomorphism of E onto G, i.e. it defines a TVS isomorphism of E/F on G. (To every element of the quotient E/F there corresponds its unique representative in G, and to each element of G its canonical image in E/F.)*

2) *The mapping of E/F onto G considered in 1) is a continuous linear mapping of E/F into E which assigns to every element in the quotient a representative element in E. Conversely, every mapping Ψ of E/F into E having these properties can be obtained in this way (clearly uniquely since we will have $G = \Psi(E/F)$).*

The proof is obvious.

COROLLARY 1 *Two topological supplements of the same space F are canonically isomorphic.*

In fact they are both canonically isomorphic to E/F.

COROLLARY 2 *If E is a Hausdorff TVS, F and G supplementary topological vector subspaces, then F and G are closed.*

In fact, they are respectively the kernels of the continuous linear mappings $1 - p$ or p.

Care should be taken, however, for, as we shall see later, even a closed vector subspace of a Banach space does not always admit a topological supplement; this causes numerous difficulties. Let us point out that if D is a differential operator with constant coefficients, *elliptic* on R^n, $n \geqslant 2$, and of strictly positive order, then it is a homomorphism of $\mathscr{E}(R^n)$ onto $\mathscr{E}(R^n)$ which does not admit a continuous right inverse (see below), i.e. its kernel has no topological supplement.

Let u be a continuous linear mapping of a TVS E into a TVS F. A continuous linear mapping v of F into E is a right inverse (resp. left inverse) of u if $u \circ v = 1$ (resp. $v \circ u = 1$).

PROPOSITION 20   *Let u be a continuous linear mapping of a TVS E into a TVS F.*

1) *u admits a continuous right inverse if and only if it is a homomorphism of E onto F whose kernel admits a topological supplement in E.*

2) *u admits a continuous left inverse if and only if it is a isomorphism of E into F, whose image admits a topological supplement in F.*

The proof is left to the reader.

Let E be a TVS, let $E_1, \ldots, E_n$ be a finite number of vector subspaces of E. E is a *topological direct sum of subspaces E_i* if the natural continuous linear mapping

$$(x_i) \mapsto \Sigma \, x_i$$

of ΠE_i into E is a topological vector isomorphism of the first onto the second. Then the projections of ΠE_i onto the factor spaces define continuous endomorphisms p_i of E, satisfying the conditions:

$$p_i^2 = p_i \quad p_i \circ p_j = 0 \quad \text{for } i \neq j$$
$$\Sigma \, p_i = 1 \qquad p_i(E) = E_i.$$

Conversely, we verify immediately that if we have a finite sequence of continuous projectors p_i into E whose products are pairwise zero and whose sum is the identity, they correspond to the decomposition of E into a topological direct sum of subspaces $E_i = p_i(E)$.

EXERCISE 1   Let E be a TVS. Every finite dimensional vector subspace of E admits a topological supplement if and only if for every $x \neq 0$ of E there exists a continuous linear form on E which does not vanish on x.

EXERCISE 2   Let F be a closed vector subspace of a Banach space E such that E/F is isomorphic (as TVS) to the space l^1 of summable sequences (with the norm

$$\| \lambda \|_1 = \sum_i | \lambda_i |).$$

Show that F admits a topological supplement (use Proposition 19. 2) and Exercise 6 of Section 9).

EXERCISE 3   Let M be a locally compact non-compact space, $\hat{M}$ its one point compactification and E a TVS. Show that $C(\hat{M}, E)$ is a topological direct sum of $C_0(M, E)$ and of the space of constant mappings of $\hat{M}$ into E.

EXERCISE 4   Let M be a topological space, the topological sum of a finite sequence of subspaces M_i. Show that $C(M)$ (see definition in Section 8) is the topological direct sum of a sequence of subspaces isomorphic to the spaces $C(M_i)$. Describe the corresponding projectors.

EXERCISE 5   Let M be a metrisable space, N a closed non-empty subspace whose boundary is separable. Show that there exists a canonical linear mapping of $C^\infty(N)$ into $C^\infty(M)$, a right inverse of the canonical linear mapping of $C^\infty(M)$ onto $C^\infty(N)$ (Kakutani). (Let (x_i) be a sequence dense in the boundary of N, let $V_{i,n}$ be the set of points $x \in M$ such that $d(x_i, x) < 1/n$ and $d(x, N) > \frac{1}{2}n$, let (V_a) be the open covering of $\complement N$ formed by the $V_{i,n}$ and by $\complement N$. Construct a family (ϕ_a) of continuous functions positive on $\complement N$, with support $\phi_a \subset V$, uniformly summable in a neighborhood of each point of $\complement N$ and having 1 for sum. Take finally, for every $f \in C^\infty(N)$, the function f on M identical to f on N and to

$$\sum_{i,n} f(x_i)\phi_{i,n}$$

in $\complement N$). Extend the result to the case $C^\infty(N, E)$:

  1) when E is a Banach space;

  2) when E is a locally convex complete space.

REMARK   Even if M is compact and N a compact subspace, it is not true in general (if M is not metrisable) that the natural mapping of $C(M)$ onto $C(N)$ admits a right inverse mapping which is linear and continuous (it is false, for example, if M is the Stone compactification of the integers, and N the complement in M of the integers).

12 Vector subspaces of finite dimension or codimension

PROPOSITION 21   *Let E be a TVS, F a vector subspace. Then the closure of F is a vector subspace.*

COROLLARY   *A hyperplane V in E is either dense or closed (i.e. if the complement of V has an interior point, V is closed).*

THEOREM 6   *Let E be a TVS, x' a non-zero linear form on E. For x' to be continuous it is necessary and sufficient that the hyperplane kernel of x' be closed.*

Proof   We must show that if the hyperplane V is closed x' is continuous. This means that setting $F = E/N$ (F is a Hausdorff TVS of dimension 1), the linear form on F obtained by passing to the quotient is continuous. But that form is of type

$$\lambda e \longmapsto \lambda,$$

where e is some element of F, so that we must prove the

LEMMA   *Let F be a Hausdorff TVS of dimension 1, e a non-zero element of F. Then the mapping $\lambda \longmapsto \lambda e$ of K onto F is a TVS isomorphism.*

Let V be a balanced neighborhood of the origin in F such that $e \notin V$. We must show that $\lambda e \longmapsto \lambda$ is continuous at the origin, that is, for $\varepsilon > 0$ there exists a neighborhood W of the origin in F such that $\lambda e \in W$ implies $|\lambda| < \varepsilon$. We may choose $W = \varepsilon V$ since $\lambda e \in W$ is the same as $e \in (\varepsilon/\lambda) V$ and, since V is balanced we would have $(\varepsilon/\lambda) V \subset V$ if we had

$$\left| \frac{\varepsilon}{\lambda} \right| \leqslant 1,$$

which would be absurd.

THEOREM 7   *Let E be a finite dimensional Hausdorff TVS, let*

$$(e_i)_{1 < i \leqslant n}$$

be a basis of E. Then the mapping

$$(\lambda_i) \rightarrow \sum \lambda_i e_i$$

of K^n onto E is a TVS isomorphism of the first space onto the second (K^n having the product topology).

This is true if $n = 1$ (by the preceding lemma); let us proceed by induction on n supposing the theorem proved for the dimension $n - 1$. The hyperplane in E generated by $e_1, \ldots, e_{n-1}$ is isomorphic to K^{n-1}, therefore complete and thus closed since E is Hausdorff. Consequently

(Theorem 6) the linear form $\Sigma\, \lambda_i e_i \longmapsto \lambda_n$ on E is continuous. Similarly, the other λ_i will be continuous, and the theorem follows.

COROLLARY 1   *Let E be a finite dimensional Hausdorff TVS. Every linear mapping of E into a TVS F is continuous. Every vector subspace of E is closed.*

For the first part take $E = K^n$. The second part is contained in

COROLLARY 2   *Let E be a Hausdorff TVS. Then every finite dimensional vector subspace of E is closed.*
    (*It is in fact complete, by Theorem 7.*)

COROLLARY 3   *Let E be a TVS, F a closed vector subspace of finite codimension. Then every linear mapping of E into a TVS G which vanishes on F is continuous. Every supplement of F is also a topological supplement.*

For E/F being finite dimensional and Hausdorff, the first assertion follows from Corollary 1. If H is a supplement of F, the corresponding projector of E onto F is continuous (since zero on F), therefore H is a topological supplement.

COROLLARY 4   *Let E be a TVS, F a closed vector subspace, G a finite dimensional subspace. Then $F + G$ is closed.*
    Let ϕ be the canonical mapping of E onto the Hausdorff space E/F, $\phi(G)$ is a closed subspace of E/F (Corollary 2), therefore its pre-image, by ϕ which is $F + G$, is closed.

13 Locally precompact TVS

THEOREM 8 (BANACH)   *Let E be a Hausdorff TVS with a precompact neighborhood of the origin. Then E is finite dimensional.*

We first prove the

LEMMA   *Let E be a TVS, F a vector subspace, V a neighborhood of the origin such that $F + V \neq E$ (for example, F closed and different from E, V bounded). Then, for $0 < k < 1$ we have $V \not\subset F + kV$.*
    Let us suppose $V \subset F + kV$, i.e. introducing $G = E/F$ and the canonical image W of V into E/F (which is still a neighborhood of the origin in G): $W \subset kW$, from this we conclude
$$W \subset kW \subset k^2W \subset \ldots \subset k^nW,$$
therefore $W \supset (1/k^n)W$ for every n. Since the union of the $(1/k^n)W$ is G (W is absorbing), we have $W = G$, i.e. $F + V = E$, which is a contra-

diction. If, for example, F is closed and different from E, and V bounded, then W is a bounded subset of a non zero Hausdorff space G, therefore $W \neq G$.

We will prove Theorem 8 by contradiction: take $0 < k < 1$; if E were infinite dimensional, we could construct by induction (Lemma) a sequence (x_i) of points of V such that, E_n being the space generated by $x_1, \ldots, x_n$, we would have $x_{n+1} \notin E_n + kV$. We would then have

$$x_i - x_j \notin kV$$

for $i \neq j$, which contradicts the precompactness of V (since kV is a neighborhood of the origin).

14 Theorem of homomorphisms, closed graph theorem

THEOREM 9 HOMOMORPHISM THEOREM (BANACH) *A linear and continuous mapping u from a metrisable and complete TVS E onto another, F, is a homomorphism.*

If we consider the quotient of E by the kernel of u, a quotient which is still a complete and metrisable TVS (Section 4, Proposition 6), the preceding theorem is equivalent to the following particular case (*Theorem of isomorphisms*): *a continuous and bijective linear mapping u from a complete and metrisable TVS E onto another one F, is an isomorphism of the first onto the second* (i.e. the inverse mapping is continuous), or equivalently, if we have on a vector space two TVS topologies for which the space is metrisable and complete, then these topologies are either not comparable or they are identical. The proof will be given in two parts.

1) Under the conditions of Theorem 9, for every neighborhood U of the origin in E, the closure of $u(U)$ is a neighborhood of the origin in F (using only the hypothesis on F). Since $u(E) = F$, the union of the

$$nu(U) = u(nU)$$

(n a positive integer) is F and since F is a *Baire space*, one of the $nu(U)$ is not rare. Therefore $u(U)$ is not rare, i.e. its closure contains a non-empty open subset W. Let $U' = U - U$, then $W - W$ is a neighborhood of the origin in F, contained in the closure of $u(U')$. Therefore the closure of the set $u(U')$ is a neighborhood of the origin, from which follows the assertion.

2) Theorem 9 is now a consequence of

LEMMA *Let u be a continuous linear mapping of a TVS E into a TVS F. We suppose that E is metrisable and complete, and that for every neigh-*

D

borhood U of the origin in E, the closure $\overline{u(U)}$ of $u(U)$ is a neighborhood of the origin in F. Then u is a homomorphism of E onto F.

(This implies therefore that F is metrisable and complete.) Precisely, let U be a neighborhood of the origin in E. We will show that every $y \in \overline{u(U)}$ is of the form ux, where $x \in U + V$, V being an arbitrary neighborhood of the origin in E (which implies the lemma). Let (U_i) be a fundamental sequence of neighborhoods of the origin in E, such that

$$U_1 = U \quad U_2 + U_2 \subset U \cap V \quad \text{and} \quad U_n + U_n \subset U_{n-1} \quad \text{for } n \geqslant 3.$$

This implies that

$$U_{n+1} + \ldots + U_{n+k} \subset U_n$$

if $n \geqslant 1$ and

$$U_1 + \ldots + U_k \subset U + V.$$

Let us construct by induction a sequence (y_i) of points of F, such that

$$y_i \in u(U_i) \quad \text{and} \quad y - (y_i + \ldots + y_n) \in \overline{u(U_{n+1})}.$$

(The possibility of induction is obvious.) Let $x_i \in U_i$ such that $ux_i = y_i$. The sequence (x_i) in E is summable since E is complete and the sum of a finite number of x_i for $i > n$ is always in U_n. Let $x = \Sigma\, x_i$ be the sum. Passing to the limit we find that $x \in \overline{V + U}$ and that

$$ux = \Sigma\, ux_i = y.$$

This proves the lemma.

COROLLARY 1 *Let E and F be two Banach spaces, U and V their unit balls, u a linear mapping of E into F such that $u(U)$ is dense in V. Then u is a metric homomorphism of E onto F.*

We have already shown that every element of $\overline{u(U)} = V$ is the image by u of an element of $\overline{U + \varepsilon U} = (1 + \varepsilon)\overline{U}$, where $\varepsilon > 0$ is given.

COROLLARY 2 *Let E be a metrisable and complete TVS, F and G two vector subspaces. For F and G to be topological supplements it is necessary and sufficient that they be closed and algebraic supplements.*

(Apply the isomorphism theorem to $F \times G$ and E.)

COROLLARY 3 *Let E and F be two TVS, E metrisable and complete, u a continuous linear mapping of E into F. If $u(E) \neq F$, $u(E)$ is meagre.*

In fact it follows from the lemma that if $u(E) \neq E$, there exists a neighborhood U of the origin in E such that $u(U)$ has no interior, but,

$$u(E) = \bigcup_i n\, u(U).$$

Another important form of the theorem of homomorphisms is

THEOREM 10, THE CLOSED GRAPH THEOREM *Let u be a linear mapping of a complete and metrisable TVS E into another, F. For u to be continuous (it is necessary and) it is sufficient that its graph be closed, or equivalently, that there is no sequence in E tending to zero such that the image sequence has a non-zero limit.*

Let H be the graph of u, i.e. the set of pairs

$$(x, ux) \in E \times F,$$

where x runs through E. If H is closed, it is a metrisable and complete TVS, and since its projection on E is a bijective continuous linear mapping of H onto E, it is a TVS isomorphism, a fact that follows from the theorem of isomorphisms. Therefore the inverse mapping is continuous, as is u obtained by composing the preceding mapping with the projection of H onto F. To say that the graph H is closed means that if (x_i) is a sequence in E converging to x, and if $(u(x_i))$ converges to y, then $y = u(x)$. Replacing x_i by $x_i - x$ we see clearly that we can restrict ourselves to the case $x = 0$.

COROLLARY *Let E and F be metrisable and complete TVS, u a linear mapping of E into F. For u to be continuous it is already sufficient that u be continuous for a Hausdorff topology on F coarser than the given topology of F.*

REMARKS Theorem 10 practically means that a linear mapping from a complete and metrisable TVS into another one, F, defined *in a natural way*, is always continuous. In fact, we do not know an *explicit* example (i.e. not using the axiom of choice) of a non-continuous linear mapping from a metrisable and complete TVS into another. However, it is easy to see that if E is a metrisable TVS of infinite dimension, there exist non-continuous linear forms on E: take a sequence (x_i) in E converging to 0, the x_i being linearly independent, and take on the vector space generated by the x_i the linear form equal to 1 on the x_i, extend this form to all of E in an arbitrary way. We should note that for the theorem of isomorphisms as in the closed graph theorem, the fact that the spaces are *complete* is essential. A counter example when F is not complete: the identity mapping of $\mathscr{E}^{(m+1)}(K)$ into $\mathscr{E}^{(m)}(K)$ (Section 10) is a bijective continuous linear mapping from the Banach space $E = \mathscr{E}^{(m+1)}(K)$ onto the subspace $F = u(E)$ of $\mathscr{E}^{(m)}(K)$, but it is not a TVS isomorphism of E onto F. A counter example when E is not complete: Let F be a complete and metrisable TVS of infinite dimension, v a non-continuous linear form on F (see above), E the space F

with the coarsest topology for which the identity mapping and v are continuous (E is then a metrisable TVS), let u be the identity mapping of E onto F: u is a bijective continuous linear mapping of E onto F whose inverse is not continuous.

EXERCISE 1   Every separable Banach space E is isomorphic to a quotient space of the space l^1 of summable sequences (isomorphisms for the normed structure). (Take a dense sequence (x_i) in the unit ball of E and show that the mapping

$$(\lambda_i) \longmapsto \sum \lambda_i x_i$$

is a metric homomorphism of l^1 onto E.) Every Banach space E (separable or not) is isomorphic to a quotient space of a space $l^1(I)$. (Choose I to be a dense subset of the unit ball of E.)

EXERCISE 2   Let M be a compact space, E a locally convex, metrisable and complete space (or a Banach space), F a closed vector subspace of E, ϕ the canonical mapping of E onto E/F. Show that the mapping $f \longmapsto \phi \circ f$ of $C(M, E)$ into $C(M, E/F)$ is a homomorphism (or a metric homomorphism) of the first space onto the second. (Use the lemma of Section 14—or Corollary 1 of Theorem 9—and the proof of Theorem 4, Section 9.) If M is only locally compact and countable to infinity, prove the analogue for $C(M, E/F)$ and $C_0(M, E/F)$.

EXERCISE 3   Let K be a compact cube of R^n, let F be a Banach space whose elements are scalar functions on K, with a topology finer than that of the simple convergence. If F contains the infinitely differentiable functions on K, there exists an integer $m \geqslant 0$ such that F also contains the m-times continuously differentiable functions on K. (Show that the identity mapping of the space $\mathscr{E}(K)$—see Section 10—into F is continuous.)

EXERCISE 4   Let E, F be a metrisable and complete TVS, u a linear mapping of E into F such that $u(E)$ is of finite codimension. Show that u is a *homomorphism*. More generally, two vector subspaces of E that are the images of metrisable and complete TVS by continuous linear mappings, which are algebraic supplements, are topological supplements.

15   The Banach–Steinhaus theorem

THEOREM 11 (BANACH–STEINHAUS)   *Let E be a metrisable and complete TVS, F any TVS, M a set of continuous linear mappings of E into F.*

For M to be equicontinuous it is necessary and sufficient that for every
x ∈ E, the set M(x) of images of x by the u ∈ M be a bounded subset of F
(i.e. that M be a subset of L(E, F) bounded for the topology of pointwise
convergence—see Section 8, Proposition 15).

The necessity is immediate; more generally,

PROPOSITION 22   *If M is an equicontinuous set of linear mappings of a*
TVS E into a TVS F, then for every bounded subset A of E, the set
$M(A) = \bigcup_{u \in M} u(A)$ *is a bounded subset of F (i.e. M is a bounded subset of*
L(E, F) for the topology of uniform convergence on the bounded subsets of
E—see Section 8, Theorem 3 and Proposition 15).

Conversely, if under the general conditions of Theorem 11 $M(x)$ is a
bounded subset of F for every $x \in E$, let us show that M is equicontinuous. It suffices to show equicontinuity at the origin, i.e. that for
every neighborhood V of the origin in F, the set

$$U = M^{-1}(V) = \bigcup_{u \in M} u^{-1}(V)$$

is a neighborhood V of the origin in E. We can assume V closed, therefore U is closed, furthermore the hypothesis on M means exactly that
the sets U are absorbing. From Baire's theorem a closed absorbing
subset U of E has a non-empty interior (see the first part of the proof of
Theorem 9) therefore $U - U$ is a neighborhood of the origin in E.
Then $M^{-1}(V - V)$ is a neighborhood of the origin in E, the conclusion
follows since as V varies $V - V$ runs through a fundamental system of
neighborhoods of the origin in F.

COROLLARY OF THE BANACH–STEINHAUS THEOREM   *Let E be a metrisable*
and complete TVS, F any Hausdorff TVS, (u_i) a sequence of con-
tinuous linear mappings of E into F, converging for every $x \in E$ towards
a limit u(x). Then (u_i) is an equicontinuous sequence, therefore u is
continuous (clearly, linear) from E into F and (u_i) tends toward u uni-
formly on every compact set.

The analogous statement for a filter with a countable basis in $L(E, F)$
is still valid since it can be reduced to the case of a sequence.

A particularly important application of the Banach–Steinhaus
theorem is the following theorem, from which the preceding one can be
obtained if F is the field of scalars:

THEOREM 12   *Let E and F be metrisable and complete TVS, G any*
TVS. For a bilinear mapping u of $E \times F$ into G to be continuous, (it is

necessary and) it is sufficient that it be continuous with respect to each variable separately. More generally, a set M of bilinear mappings of $E \times F$ into G is equicontinuous, (if and) only if the $u \in M$ are separately continuous, and for every $(x, y) \in E \times F$, the set $M(x, y)$ of images $u(x, y)$ for $u \in M$, is a bounded subset of G.

The necessity is immediate. The sufficiency follows from

LEMMA *Let E be a metrisable and complete TVS, T a metrisable topological space, G a TVS, M a set of mappings*

$$(x, t) \longmapsto u(x, t)$$

of $E \times T$ into G, linear with respect to x. For M to be equicontinuous it is necessary and sufficient that the following three conditions be satisfied:

1) *the $u \in M$ are continuous with respect to x;*

2) *the set M is equicontinuous with respect to t, i.e. for every $x_0 \in E$, the set of mappings $t \longmapsto u(x_0, t)$ from T into G is equicontinuous;*

3) *M is bounded for the topology of the simple convergence, i.e. for every $(x, t) \in E \times T$, the set $M(x, t)$ of $u(x, t)$ for $u \in M$ is a bounded subset of G.*

(This lemma was found in a manuscript of N. Bourbaki.)

The lemma implies Theorem 12 since the conditions of Theorem 12 imply the conditions in the lemma. Condition 2) of the lemma will be verified using the Banach–Steinhaus theorem). To prove the lemma, we return to the case where T is compact, and use the following fact (immediately verifiable): Let P be a metrisable space, M a set of mappings from P into a uniform space G: for M to be equicontinuous at $p \in P$, it is necessary and sufficient that for every sequence (p_n) tending towards p, the set of restrictions of $u \in M$ to the set of (p_n) and p be equicontinuous at p. When T is compact, Hypotheses 2) and 3) imply that for every $x_0 \in E$, the set of $u(x_0, t)$ for $u \in M$ and $t \in T$ is a bounded subset of G (by using the compactness). But by virtue of Hypothesis 1) and the Banach–Steinhaus theorem, this implies that the set of all mappings $x \longmapsto u(x, t)$, when u varies in M and t in T, is an *equicontinuous* set of mappings of E into G. If W is a neighborhood of the origin in G and $(x_0, t_0) \in E \times T$, there exists a neighborhood U of x_0 such that $u(x, t) - u(x_0, t) \in W$ as soon as $x \in U$, for any $t \in T$ and $u \in M$. On the other hand by virtue of Hypothesis 2) there exists a neighborhood V of t_0 in T such that

$$u(x_0, t) - u(x_0, t_0) \in W$$

for every $t \in V$, and any $u \in M$. We have then

$$u(x, t) - u(x_0, t_0) \in W + V$$

for $x \in U$, $t \in V$ for any $u \in M$. Since W was an arbitrary neighborhood of the origin in G, it follows that M is equicontinuous.

EXERCISE   Let E be a topological space, F a subset of E; F is meagre with respect to E at a point $x \in E$ if there exists a neighborhood V of x such that $V \cap F$ is a meagre subset of E.

a) Show that if E is a Baire space and if F is not meagre at any point $x \in F$, then F is a Baire space.

b) Show that if F is a Baire space dense in E, then E is a Baire space and F is meagre at x, at no point $x \in F$ (with respect to E). (Therefore if E is a dense topological subspace of E, F is a Baire space if and only if E is a Baire space and F is not meagre with respect to E at any point $x \in F$.)

c) Let E be a TVS, F a dense vector subspace. F is a Baire space if and only if E is a Baire space and F is not meagre in E. In particular, if F is a metrisable TVS, F is a Baire space if and only if F is not meagre in its completion.

d) Let E be a TVS. E is a Baire space if and only if E is not the union of a sequence of closed sets having no interior.

CHAPTER 2

The general duality theorems on locally convex spaces

1 Introduction

As the title indicates, we group in this Chapter all duality properties which follow directly from the Hahn–Banach theorem, insofar as they concern the *most general* locally convex spaces. Particular categories of spaces will be studied in the next chapters.

In contrast to the preceding chapter, the particular structure of the field of real numbers and in particular its order structure will be important. Since the field of real numbers $\mathbf{R}$ is a subfield of the field of complex numbers $\mathbf{C}$, every vector space E over $\mathbf{C}$ has also the structure of a vector space E_0 over $\mathbf{R}$. For a topology on E there is no difference between its compatibility with the vector structure over $\mathbf{R}$ or over $\mathbf{C}$; if E is a complex TVS, we can consider its associated real TVS E_0.

We will call the space E' of *continuous* linear forms on E the *dual* space (or *dual*) of the TVS E (real or complex). We then have:

PROPOSITION 1 *Let E be a complex TVS, E_0 the associated real TVS. If to every continuous linear form x' on E we assign the function*

$$x \longmapsto \mathscr{R}\langle x, x' \rangle$$

on $E = E_0$, we get a bijective mapping of the dual E' onto the dual $(E_0)'$ of E_0, which is an isomorphism for the real vector structure of E' and $(E_0)'$. If x' is the complex linear form corresponding to the continuous real linear form y', then the kernel of x' is $V \cap iV$, where V is the kernel of y'.

For the first part, we notice that $\mathscr{R}\langle x, x' \rangle$ is a real continuous linear form on E, call it y'; also, that x' can be expressed with the aid of y' by

$$\langle x, x' \rangle = \langle x, y' \rangle - i\langle ix, y' \rangle$$

and that conversely, if y' is given real continuous linear form on E, the preceding formula defines x' as a complex continuous linear form. The characterization of the kernel of x' is obvious.

2 Convex sets, disked sets

DEFINITION 1   *Let A be a subset of a vector space E. The subset A is convex if together with two points x and y, A contains the segment that joins them.*

Recall that the segment joining the points x and y is the set of points $\lambda x + (1 - \lambda)y$, where $0 \leqslant \lambda \leqslant 1$. A set reduced to a point, a linear sub-variety, a segment, are examples of convex sets. If F is a vector subspace of E containing A, it is equivalent to say that A is convex in F or in E.

PROPOSITION 2   *Let E be a real one-dimensional vector space. The convex subsets of E are the intervals of E (finite or infinite, closed or open or semi-open).*

The proof is immediate.

PROPOSITION 3   *Every intersection, every increasingly directed union of convex sets is convex; the direct or inverse image of a convex set by a linear mapping is convex; if E_i are vector spaces and $A_i \subset E_i$ convex subsets, then the product of the A_i is a convex subset of the product of the E_i; the sum of a family of convex sets A_i of a vector space E is convex; a translate, a homothetic image of a convex set is convex.*

The proof is trivial: the first five properties (closure) are easily verified; the case of the sum reduces to the case of a finite sum, then we consider $\Sigma\, A_i$ as the image of the convex set $\Pi\, A_i$ by the linear mapping $(x_i) \mapsto \Sigma\, x_i$ of E^n into E. In particular, if A is convex, $A + x$ is convex for every $x \in E$, since $\{x\}$ is convex.

Since E is convex and an intersection of convex sets is convex, there exists a smallest convex set containing a given set A, called the *convex hull* of A. We denote it by $\Gamma(A)$.

PROPOSITION 4   *Let A be a subset of a vector space. Then its convex hull $\Gamma(A)$ is the set of sums*

$$\sum_{1 \leqslant i \leqslant n} \lambda_i x_i,$$

where the x_i are elements of A and the λ_i are positive scalars of sum 1.

In order to prove Proposition 4, it is sufficient to show that the set of these sums $\Sigma\, \lambda_i x_i$ is convex (immediate), and that such a sum is contained in every convex set B containing A. This is trivial for a sum of 1 or 2 terms, and can be verified by induction in the general case.

DEFINITION 2 *Let A be a subset of a vector space E. Then A is called a disked set or simply a disk, if together with two points x and y it also contains all the points $\lambda x + \mu y$ where λ and μ are two scalars such that $|\lambda| + |\mu| \geqslant 1$.*

Such a set is a fortiori convex, and furthermore balanced (see Chapter 1, Proposition 2), that is, closed under homothetics of norm $\leqslant 1$. The converse is true and furthermore:

PROPOSITION 5 *Let A be a subset of a vector space E. The following conditions are equivalent:*

a) *A is disked.*

b) *A is convex and balanced.*

c) *A is convex and $\lambda A \subset A$ for $|\lambda| = 1$.*

It suffices to show that c) implies b) and that b) implies a). Supposing c) to be verified by us prove that A is circled, i.e. $\lambda A \subset A$ for all λ with $|\lambda| \leqslant 1$. Writing

$$\lambda = |\lambda| \frac{\lambda}{|\lambda|}$$

when λ is not zero we see that it is sufficient to consider the case in which $0 \leqslant \lambda \leqslant 1$. Now from c) we see that $-x \in A$ whenever $x \in A$. Thus from the convexity it follows that the whole segment $(-x, x)$ is contained in A and that in particular the points λx with $0 \leqslant \lambda \leqslant 1$ are in A. Supposing now b) to be verified let us prove a), i.e. $\lambda x + \mu y \in A$ whenever x and y are in A and $|\lambda| + |\mu| \geqslant 1$. Writing

$$\lambda = |\lambda| \frac{\lambda}{|\lambda|}, \qquad \mu = |\mu| \frac{\mu}{|\mu|}$$

we see again that we may suppose $\lambda > 0$, $\mu > 0$ and furthermore $\lambda + \mu > 0$. It suffices then to write

$$\lambda x + \mu y = (\lambda + \mu)\left(\frac{\lambda}{\lambda + \mu}x + \frac{\mu}{\lambda + \mu}y\right).$$

The *stability properties* stated in Proposition 3 are all true for disked sets except that a translate of a disked set is not in general disked. We can show this directly as in Proposition 3 or as a corollary to Proposition 3, using criterion c) of Proposition 5. Since a vector subspace of E and in particular E itself, is clearly disked, we can conclude that there exists a smallest disked subset containing a given subset A; we call it the *disked hull* of A. We show as for Proposition 5 that the disked hull of A is the set of sums $\Sigma \, \lambda_i x_i$, where the x_i are points of A and the λ_i scalars such that $\Sigma \, | \, \lambda_i \, | \leqslant 1$.

In the case where the field of scalars is **R**, only $+1$ and -1 have norm 1, therefore (Proposition 5, c)) the disked sets are the symmetric convex sets (A is *symmetric* if $A = -A$).

3 Convex cones and ordered vector spaces

Even though the elementary theory which follows is not necessary to the development of the theory of TVS, we present it because of its interest in applications.

DEFINITION 3   *A subset A of a vector space E is called a cone if $\lambda A \subset A$ for every $\lambda > 0$.*

A may or may not contain the origin; the definition means also that A is the union of "open" half-rays from the origin O. The empty set or a vector subspace are particular cones.

PROPOSITION 6   *Let A be a subset of a vector space. A is a convex cone if and only if*
$$A + A \subset A \quad \text{and} \quad \lambda A \subset A$$
for $\lambda > 0$.

For the necessity, write $x + y = 2(\frac{1}{2}x + \frac{1}{2}y)$, which shows that if $x, y \in A$ then $x + y \in A$. The sufficiency is still more trivial.

We have furthermore for convex cones stability properties exactly identical to those of Proposition 3, except that a translate of a convex cone is not in general a convex cone; these properties can be verified directly or as a corollary to Proposition 3. If A is any subset of E, there exists a smallest convex cone containing A called the *convex cone generated by A*. It is clear that the convex cone generated by A is the set of sums $\Sigma \, \lambda_i x_i$ with $x_i \in A$ and the λ_i scalars > 0.

DEFINITION 4   *Let E be a vector space. A pre-order structure on E is said to be compatible with the vector structure of E if it is invariant by translations and by strictly positive homothetics. A vector space E with a (pre-) order structure compatible with the vector structure is called a (pre-) ordered vector space.*

The axioms mean therefore that $x \leqslant y$ implies $x + z \leqslant y + z$ for every $z \in E$ and $\lambda x \leqslant \lambda y$ for every $\lambda > 0$. From this we infer that $x_1 \leqslant y_1$ and $x_2 \leqslant y_2$ imply $x_1 + x_2 \leqslant y_1 + y_2$. In particular, $x \leqslant y$ is equivalent to $y - x \geqslant 0$, therefore the pre-order is known once we know the set P of elements $\geqslant 0$ (called *positive elements*). From Proposition 6, we see that P is a convex cone, clearly containing the origin. Conversely, let P be a convex cone in E containing zero; we verify immediately that the relation $y - x \in P$ defined on E is a pre-order relation (reflexive because $O \in P$, transitive since $P + P \subset P$), compatible with the vector structure ($\lambda P \subset P$ for $\lambda > 0$). We have shown the first part of:

PROPOSITION 7

1) *Let E be a vector space. There is a bijective correspondence between the pre-order relations on E compatible with the vector structure, and the convex cones in E containing the origin. To a pre-order relation compatible with the vector structure there corresponds the cone P of positive elements of E, and to P there corresponds the pre-order defined by: $x \leqslant y$ if and only if $y - x \in P$.*

2) *A pre-ordered vector space E is ordered if and only if*
$$P \cap (-P) = \{0\};$$
it is increasingly directed (for all $x, y \in E$, there exists a z such that $z \geqslant x, z \geqslant y$). ($P$ stands for the cone of positive elements.)

Proof of the second part is left to the reader.

4 Correspondence between semi-norms and absorbing disks. Characterization of locally convex spaces

THEOREM 1   *Let E be a vector space, p a semi-norm on E. The set U of all $x \in E$ such that $p(x) \leqslant 1$ (the unit ball associated with p) is an absorbing disk and p is known once we know U, by virtue of the formula:*
$$p(x) = \inf_{\substack{\lambda > 0 \\ x \in \lambda U.}} \lambda$$

Conversely, if U is an absorbing disk in E, the preceding formula (which defines a positive function p, U being absorbing) defines a semi-norm p on E (called the gauge of U). Its unit ball is U with the addition of the ends of intervals defined by the intersection of U with the real rays passing through the origin.

The first part is trivial. For the converse, it is clear that
$$p(\lambda x) = |\lambda| \, p(x)$$
for every λ (since U is balanced). In order to show that
$$p(x + y) \leqslant p(x) + p(y)$$

we note that $p(x) < \lambda$ and $p(y) < \mu$ imply $p(x + y) \leqslant \lambda + \mu$, since

$$\lambda U + \mu U = (\lambda + \mu)\left(\frac{\lambda}{\lambda + \mu}U + \frac{\mu}{\lambda + \mu}U\right) \subset (\lambda + \mu)U$$

(U being convex). The characterization of the unit ball associated with p is again trivial. We have then a bijective correspondence between semi-norms on E on the one hand and on the other hand the absorbing disks whose intersection with every real homogeneous ray is closed. In particular:

COROLLARY 1   *Let E be a TVS. Then the continuous semi-norms on E correspond bijectively to the disked and closed neighborhoods of the origin O in E.*

THEOREM 2   *Let E be a TVS. The following conditions are equivalent:*

a) *E is locally convex (Chapter 1, Section 6, Definition 4);*

b) *E admits a fundamental system of disked neighborhoods of O;*

c) *E admits a fundamental system of convex neighborhoods of O.*

The equivalence of a) and b) results from Corollary 1, since saying that E is locally convex is equivalent to saying that the unit balls associated with the continuous semi-norms on E form a fundamental system of neighborhoods of the origin. It suffices to show then that c) implies b), and for this we note that if V is a convex neighborhood of O, then $\bigcap\limits_{|\lambda|=1} \lambda V$ is a *neighborhood* of O which is *convex* (intersection of convex sets), closed under homothetics of norm 1, therefore disked (Proposition 5).

Let E be a vector space. The least upper bound of all locally convex topologies on E is a locally convex topology, which is the finest locally convex topology on E. A convex subset V of E is a neighborhood of O if and only if it is absorbing. The necessity is trivial, for the sufficiency we may assume V to be disked (if not we can replace it by $\bigcap\limits_{|\lambda|=1} \lambda V$); then V contains the interior of the unit ball defined by the semi-norm p, the gauge of V, therefore it is a neighborhood of O for the semi-normed topology defined by p; the assertion follows. A point $x \in E$ is an *internal point of a set V* if $V - x$ is absorbing; we see then that if V is convex it is equivalent to saying that x is interior to V for the finest locally convex topology, which reduces a notion of purely algebraic nature to a topological one.

5 Convex sets in TVS

PROPOSITION 8   *Let E be a TVS. The closure of a convex set (or a disk, or a convex cone) is convex (or a disk or a convex cone).*

The closure of the convex hull $\Gamma(A)$ of a set A is called the closed convex hull of A, written $\bar{\Gamma}(A)$. We define similarly the closed disked hull of A (or closed disk generated by A), and the closed convex cone generated by A.

PROPOSITION 9   *Let E be a TVS, A a convex subset of E. Then $\mathring{A}$ is convex and if it is non-empty, its closure is identical to the closure of A. If x_0 is an interior point of A, then the interior of A (or the closure of A, or the boundary of A) is the union of open intervals (or closed intervals, or finite extremities of intervals) defined by the intersection of A with the real rays passing through x_0.*

The proof follows immediately from:

LEMMA   *Let A be a convex subset of a TVS E, x an interior point of A, y a point of the closure of A; then the segment (x, y) minus y is interior to A.*

Let $0 < \lambda \leqslant 1$, let us show that $\lambda x + (1 - \lambda)y$ is interior to A. Let $U = \mathring{A}$. Consider the set of $z \in E$ such that

$$\lambda x + (1 - \lambda)y \in \lambda U + (1 - \lambda)z$$

i.e.   the set $y + \dfrac{\lambda}{1 - \lambda}(x - U)$;

it is an *open* set containing y (since $x \in U$) therefore (y in the closure of A) it contains $z \in A$. But since $U \subset A$, $\lambda U + (1 - \lambda)z$ is an open set contained in A and containing $\lambda x + (1 - \lambda)y$, the conclusion follows.

COROLLARY   *The interior of a disk is a disk, the interior of a convex cone is a convex cone. Let p be a continuous semi-norm on E, U the associated unit ball. U is closed, its interior is the set of x such that $p(x) < 1$, its boundary the set of x such that $p(x) = 1$.*

EXERCISE 1   Let A be a convex set in a finite dimensional vector space E; if the affine linear variety (manifold) generated by A is E, then A has a non-empty interior.

EXERCISE 2   Let A be a subset of a real n-dimensional vector space E. Show that the convex hull of A is the set of sums with $n + 1$ terms

$\Sigma \lambda_i x_i$ where $x_i \in A$, $\lambda_i \geqslant 0$, $\Sigma \lambda_i = 1$. From this conclude that if A is compact, $\Gamma(A)$ is compact.

EXERCISE 3   Let A be a convex compact subset of a real n-dimensional vector space E, and suppose that A has a non-empty interior. Show that A is homeomorphic to the unit ball of R^n.

EXERCISE 4
a) Let A and B be two convex subsets of a Hausdorff TVS E. Show that the convex hull of $A \cup B$ is the set of sums $\lambda x + (1 - \lambda)y$ with $x \in A$, $y \in B$ $0 \leqslant \lambda \leqslant 1$.
COROLLARY   If A and B are compact convex sets, then $\Gamma(A \cup B)$ is compact.

b) Let A be a convex subset of E containing the origin, show that its disked hull is contained in $A - A$ in the case of real scalars, contained in $(A - A) + i(A - A)$ in the case of complex scalars.
COROLLARY   If A is convex, relatively compact, its disked hull is relatively compact.

EXERCISE 5   Let E be a locally convex space. Show that the closed disked hull of a precompact subset of E is precompact. Consider the case where every closed bounded set in E is complete.

EXERCISE 6   Let E be a TVS, A a convex subset of E, V a vector subspace, then $\overline{\Gamma}(A \cup V) = \overline{A + V}$. From this conclude that if E is Hausdorff, A is convex *compact*, V a *closed* subspace, then

$$\overline{\Gamma}(A \cup V) = A + V.$$

(Show that if A and B are two subsets of E, A compact, B closed, then $A + B$ is closed.)

6 The Hahn–Banach theorem

In the next two sections, a *hyperplane* is a not necessarily homogeneous affine linear subvariety of codimension 1. If E is a TVS, we have seen in Chapter 1, Section 12 (Proposition 21 and Theorem 6) that a hyperplane V in E is closed if and only if its complement has at least an interior point or furthermore if and only if V being given by the equation $x \in V \Leftrightarrow \langle x, x' \rangle = \alpha$, the linear form x' on E is *continuous*. Notice that there is therefore a bijective correspondence between *non*-homogeneous closed hyperplanes of E and the continuous non-zero linear forms on E, attaching to the linear form x' the hyperplane of equation

$\langle x, x' \rangle = 1$. To say that V does not meet a set A is to say that x' does not have the value 1 in its range when its domain is A; if A is disked, that means also that $|\langle x, x' \rangle| < 1$ for $x \in A$. If A is furthermore open (therefore the interior of the unit ball associated to a continuous semi-norm p on E) it means also that

$$|\langle x, x' \rangle| \leqslant 1$$

for $x \in A$, i.e. $|\langle x, x' \rangle| \leqslant 1$ for $p(x) \leqslant 1$, that is to say x' considered as a linear form on the space E semi-normed by p is of norm $\leqslant 1$ (Chapter I, Section 5).

HAHN–BANACH THEOREM I *Let E be a real or complex TVS, U a convex, open and non-empty subset of E, V a linear sub-variety of E not meeting U. Then there exists a closed hyperplane containing V and not meeting U.*

Proof Suppose the origin is contained in V (translate if necessary), that is, V is a vector subspace. Furthermore we can restrict ourselves to the case of real scalars since in the complex case if W is a real hyperplane passing through V not meeting U, then $W \cap iW$ is a complex hyperplane satisfying the same conditions. Let M be the set of subspaces of E containing V and not meeting U. Ordered by inclusion M is clearly inductive, let W be a maximal element, it suffices to show that W is a hyperplane which will necessarily be closed since U is open and contained in $\complement W$. We will show that if W were of codimension $\geqslant 2$, W would not be maximal. In fact, $F = E/W$ would be of dimension $\geqslant 2$ and the image U' of U by the canonical mapping ϕ of E onto F would be an open convex subset of F not containing O; if we construct a homogeneous ray D in F not meeting U' the result will follow since $\phi^{-1}(D)$ will be in M, will contain W and will be different from W. We then find a homogeneous ray D in F not meeting a given convex open subset U' not containing the origin. Since $\dim F \geqslant 2$, we are led back to the case where F is two-dimensional and we can assume F Hausdorff. Furthermore, we verify that $C = \bigcup_{\lambda > 0} \lambda U'$ is an open convex cone not containing the origin and containing U', so that we can replace U' by C. Since the complement of $\{0\}$ is connected and clearly distinct from C, C has at least a boundary point x in its complement. C being open, and a cone, the λx with $\lambda \geqslant 0$ are also boundary points, therefore not contained in C. Also, if $\lambda \geqslant 0$, λx cannot be an element of U, since if it were, all the points of the segment $(\lambda x, x)$ minus x would be in U (Section 4, Lemma), a contradiction. Thus, the homogeneous ray passing through x satisfies the requirement.

Suppose now that the set U in the preceding theorem is disked. It therefore contains the origin, and consequently the linear variety V is not homogeneous; thus, V is in the vector space F generated by V a non-homogeneous hyperplane of F. Using the remarks at the beginning of this Section, we find: every linear form x' on a subspace F such that $|\langle x, x' \rangle| < 1$ for $x \in U \cap F$ is the restriction of a linear form y' on E such that $|\langle x, y' \rangle| < 1$ for $x \in U$. In other words:

HAHN–BANACH THEOREM II   *Let E be a TVS, p a continuous semi-norm, F a vector subspace of E and x' a linear form on F of norm $\leqslant 1$ (for the semi-norm induced by p). Then x' is the restriction to F of a linear form y' on E, of norm $\leqslant 1$ (for the semi-norm p on E).*

In the statement we have omitted to say that x' and y' are continuous since they are automatically so. It is clear that the topology of E does not in fact form part of the statement, only that of the semi-norm p.

COROLLARY 1   *Let E be a locally convex vector space, F a vector subspace. Then every continuous linear form on F is the restriction of a continuous linear form on E. Every equicontinuous set of linear forms on F is the set of restrictions of an equicontinuous set of linear forms on E.*

It is sufficient to notice that a set A of linear forms on a locally convex space (F for example) is equicontinuous if and only if there exists a neighborhood U of O such that every $x' \in A$ is bounded by 1 (in the sense of the absolute value) on U; or else, that there exists a continuous semi-norm p on F such that all the $x' \in A$ are of norm $\leqslant 1$ (when E is semi-normed by p). In particular:

COROLLARY 2   *Let E be a Hausdorff locally convex space. Then for every non-zero $x \in E$, there exists $x' \in E'$ such that*

$$\langle x, x' \rangle = 1.$$

It is sufficient to consider the linear form $\lambda x \longmapsto \lambda$ on the ray F generated by x, which is continuous since E, therefore F, is Hausdorff, and to apply Corollary 1. More generally we show:

COROLLARY 3   *Let E be a Hausdorff locally convex space, let $x_1, \ldots, x_n$ be linearly independent elements of E, and $C_1, \ldots, C_n$ any scalars. Then, there exists a continuous linear form x' on E such that $\langle x_i, x' \rangle = C_i$ for every i.*

In particular, let x_i' be a continuous linear form on E such that $\langle x_j, x_i' \rangle = \delta_{ij}$ (the Kronecker delta function) for every j; then the operator

$$\sum x'_i \otimes x_i$$

in E is a continuous projection of E onto the vector space generated by the x_i. From this we get (Chapter 1, Section 11):

COROLLARY 4   *Let E be a Hausdorff locally convex space. Every vector subspace of finite dimension of E admits a topological supplement.*

EXERCISE   Let E be a TVS, P a convex cone in E containing the origin and having a non-empty interior (it is then the set of positive elements of E for a certain preordered, directed vector space structure on E, (see Section 3)); let F be a vector subspace of E meeting the interior of P. Show that every "positive" linear form on E, i.e. taking positive values on P, is continuous, and that every positive linear form on F is the restriction of a positive linear form on E.

7 Separation of convex sets. Characterization of the closure of a convex set

In this Section we work with *real* vector spaces only.

Let E be a TVS of dimension $\geqslant 1$, x' a continuous linear form on E, and α a scalar. Then the set M of $x \in E$ such that $\langle x, x' \rangle \leqslant \alpha$ is a closed and convex set whose interior $\overset{\circ}{M}$ is the set of points $x \in E$ such that $\langle x, x' \rangle < \alpha$ and whose boundary is the hyperplane V of equation $\langle x, x' \rangle = \alpha$. A set such as M is called a *closed half-space*, the interior of such a set is called an *open half-space*. Since the inequalities $\langle x, x' \rangle \geqslant \alpha$ and $\langle x, x' \rangle > \alpha$ can be written $\langle x, -x' \rangle \leqslant -\alpha$ and $\langle x, -x' \rangle < -\alpha$, they define also a closed half-space M_1, or an open half-space $\overset{\circ}{M}_1$ respectively. We verify that there exist exactly two closed half-spaces whose boundary is the closed hyperplane V, namely M and M_1; we call them the *closed half-spaces defined by the (closed) hyperplane V*. Their intersection is V, while the corresponding open half-spaces do not meet.

A closed hyperplane V *separates* (*resp. strictly separates*) two subsets A and B of E if A is contained in one closed (*resp.* open) half-space and B in the other half-space defined by the hyperplane V. This means therefore that by putting V in the form of an equation $\langle x, x' \rangle = \alpha$, x' takes values $\leqslant \alpha$ on A and values $\geqslant \alpha$ on B (*resp.* values $< \alpha$ on A and values $> \alpha$ on B), or vice-versa. Alternatively, we can say that the point α (in R) separates (*resp.* strictly separates) the sets $x'(A)$ and $x'(B)$. If x' is given, the existence of an α such that the hyperplane $\langle x, x' \rangle = \alpha$ separates A and B is possible if and only if there exists an α separating $x'(A)$ and $x'(B)$. In particular, if A and B are convex, $x'(A)$ and $x'(B)$ are convex, therefore they are intervals (Proposition 2), and clearly it is *sufficient* that these intervals do not meet, i.e. that x' does not vanish on $A - B$, that is, that the homogeneous hyperplane W defined by x'

does not meet $A - B$. Applying the Hahn–Banach theorem I, we find the first part of

PROPOSITION 10 *Let E be a locally convex TVS, let A and B be two disjoint convex subsets of E.*

1) *If either A or B is open, there exists a closed hyperplane separating A and B.*

2) *If A is closed, B compact, there exists a closed hyperplane strictly separating A and B.*

The second part can be obtained from the first: it results from the hypothesis that there exists a neighborhood U of O such that $A + U$ and $B + U$ do not meet (this follows from a well-known property of compact sets). We can choose U convex since E is locally convex. Then $A + U$ and $B + U$ are disjoint open sets to which we apply 1).

Applying 2) to the case of a closed convex set and to a set reduced to a point, we find

THEOREM 3 *Let A be a convex subset of a locally convex space E. A is closed if and only if it is the intersection of a family of closed half-spaces.*

From this we deduce the apparently more general result:

COROLLARY 1 *Let E be a locally convex space, A a subset of E. Then the closed convex hull $\bar{\Gamma}(A)$ is the intersection of the closed half-spaces containing A.*

To say that $x_0 \in E$ belongs to $\Gamma(A)$ means therefore that for every $x' \in E'$ and every scalar α such that $\langle x, x' \rangle \leqslant \alpha$ for $x \in A$, we have $\langle x_0, x' \rangle \leqslant \alpha$.

COROLLARY 2 *Let E be a locally convex space, A a subset of E. Then the closed vector space generated by A is the intersection of closed hyperplanes containing A.*

We can prove that if V is a closed vector space and $x \in \complement V$, there exists a closed hyperplane containing V and not x, which is a result of Theorem 3 (obtained more rapidly from a direct application of the Hahn–Banach theorem I, in which we choose a neighborhood U of x which does not meet V). Explicitly, an $x \in E$ belongs to the closed vector space generated by A if and only if every continuous linear form vanishing on A vanishes on x.

DEFINITION 5 *A subset A of a TVS E is total if the closed vector space generated by A is E.*

Corollary 2 above gives:

COROLLARY 3 *Let E be a locally convex space, A a subset of E. For A to be total it is necessary and sufficient that every continuous linear form on E which vanishes on A, is identically zero.*

The following proposition is sometimes useful:

PROPOSITION 11 *Let E be a TVS. For E to be separable it is necessary and sufficient that there exists a countable total subset of E.*

The condition is clearly necessary; conversely, let A be a countable total subset of E. Since the set of *rational* linear combinations of elements of A is dense in V and countable, our assertion follows.

Notice that the Hahn–Banach theorem II and Theorem 3 above are the most convenient forms of the Hahn–Banach theorem I (in fact, they are the most general). The next two Sections will present Theorem 3 in a useful form.

EXERCISE Prove the Hahn–Banach theorem I using either Proposition 10 or Theorem 3. (Recall that in the statement of the Hahn–Banach theorem I, in the case where V is a point x; if starting from Theorem 3, distinguish first of all the case $x \notin \bar{U}$ and treat the case where x is a boundary point of U passing to the limit by supposing $O \in U$; compare with the exercise of Section 8.)

8 Dual system, weak topology

DEFINITION 5 *A dual system is a pair of vector spaces E, E′ with a bilinear form on their product E × E′.*

The value of this form on (x, x') is denoted by $\langle x, x' \rangle$. We also say that the spaces E, E' are *set in duality* by the form $\langle x, x' \rangle$. A duality between E, E' is equivalent to the existence of a linear mapping from E' into the algebraic dual E^* of E which assigns to $x' \in E'$ the form $x \mapsto \langle x, x' \rangle$ on E; this is equivalent to the existence of a linear mapping from E into the algebraic dual E'^* of E'. The duality between E and E' is said to be *separated in E′* (*resp. in E*) if the mapping from E' into E^* (*resp.* from E into E'^*) which corresponds to the duality, is injective. Ordinarily we identify E' with a vector subspace of E^* (or E with a vector subspace of E'^*). Conversely, a vector space E and a vector subspace E' of its algebraic dual E^* define a dual system (E, E'), separated in E'. It is separated in E if and only if every $x \in E$ on which all the $x' \in E'$ vanish is precisely the O element. The duality (E, E') is *separated* if it is separated in E and in E'.

Important example: Let E be a locally convex space (henceforward abbreviated to LCTVS), E' its dual, then E and E' form a dual system separated in E'. It is separated in E if and only if E is Hausdorff (use the Hahn–Banach theorem II, Corollary 2).

DEFINITION 6   *Let (E, E') be a dual system. The weak topology on E is the coarsest topology on E for which the linear forms $x \longmapsto \langle x, x' \rangle$, $x' \in E'$, are continuous. We define symmetrically the weak topology on E'. They are denoted by $\sigma(E, E')$ and $\sigma(E', E)$.*

In particular, if E is an LCTVS and E' its dual, then (E, E') is a dual system to which corresponds a weak topology on E and a weak topology on E'; by "the weak topology" on E or E' we shall mean, unless otherwise specified, the weak topologies defined above.

E and E' are LCTVS with these weak topologies. E is Hausdorff for the weak topology if and only if the duality between E and E' is separated in E. (We have the same criterion for E' and from now on we will not repeat the corresponding statements for E'.) We notice that the weak topology of E depends only on the image of E' in E^*, therefore it is also the weak topology defined by the subspace of E^*.

A subset of E is said to be *weakly closed* (or *weakly open, weakly compact*, etc.) if it is closed (or open, compact, etc.) for the weak topology. Care should be taken in the case where E is a LCTVS, as the notion of a weakly closed or weakly open subset of E is *more* restrictive than that of a closed or open subset; it is the contrary for weakly compact or compact.

A fundamental system of neighborhoods of the origin in E is formed by sets of $x \in E$ such that $|\langle x, x'_i \rangle| \leqslant 1$ for every i, where (x'_i) is a finite set of elements of E'. Each V contains the vector subspace H formed by the x such that $\langle x, x_i \rangle = 0$ for every i, from which we conclude that every linear form on E bounded above on V, is zero on H, therefore it is a linear combination of linear forms defined by the x'_i. Whence:

PROPOSITION 12   *Let (E, E') be a dual system. Then the weakly continuous linear forms on E are exactly those defined by the $x' \in E'$.*

Clearly, the analogue is true on E' by considerations of symmetry.

COROLLARY   *Let E be an LCTVS. The linear forms which are continuous for the initial topology or the weak topology are the same.*

THEOREM 4   *Let E be an LCTVS, A a convex subset of E. A is closed if and only if A is weakly closed.*

First we can reduce the proof to the case of real scalars, for it is easily verified with the aid of Proposition 1 that if E is a complete LCTVS, its weak topology is identical to the weak topology of the associated real TVS. If A is closed, A is the intersection of a family of closed half-spaces (Theorem 3); since a closed half-space is defined by an equation $\langle x, x' \rangle \leqslant \alpha$ with $x' \in E'$, it is also a weakly closed set, therefore A is weakly closed.

COROLLARY   *Let E be an LCTVS, A a subset of E. Then the closed convex hull of A is identical to its weakly closed convex hull. In particular, if A is convex its closure is identical to its weak closure.*

THEOREM 4'   *Let E be a LCTVS, A a subset of E. A is bounded if and only if A is weakly bounded.*

It is sufficient to show that if A is weakly bounded, it is already bounded. If E is normed, then the natural mapping of E into the dual of E', where E' is considered as a normed space (see Chapter 1, Section 5), is a metric isomorphism (by the Hahn–Banach theorem II, taking F to be the ray generated by an $x \in E$). The weak topology of E is then that induced by the simply convergent topology in the dual of E'. By the Banach–Steinhaus theorem (Chapter 1, Section 15, Theorem 11) A is then an equicontinuous set of linear forms on E', since it is bounded for the simple convergence, therefore A is also bounded for the norm of E. In the case where E is any LCTVS, we consider the topology of E as the coarsest for which given linear mappings u_i from E into normed spaces E_i are continuous (Chapter 1, Section 6, Proposition 9) and to prove that A is bounded reduces to proving that the $u_i(A)$ are bounded subsets of E_i (Chapter 1, Section 7, Proposition 11). But the u_i are clearly continuous for the weak topologies of E and E_i, therefore the $u_i(A)$ are weakly bounded subsets of E_i. We are thus led back to the particular case first considered. Owing to Theorem 4', there is no ambiguity, if (E, E') is a dual system, in saying simply "bounded subsets", instead of "weakly bounded subsets" in the space E or E'.

Proposition 12, and even more Theorems 4 and 4', reduce many questions relative to an LCTVS E to considerations of its weak topology. It is therefore of interest sometimes to substitute this last topology for the initial topology in order to reason in terms of duality. Care should be taken as the weak topology of a LCTVS E seldom makes E a complete space. Precisely:

PROPOSITION 13 *Let (E, E') be a separated dual system. Then the algebraic dual E'^* of E', equipped with the weak topology $\sigma(E'^*, E')$, is the completion of E in the weak topology.*

In fact, E with the weak topology is identified with a topological vector subspace of E'^*, then it suffices to show that E'^* is complete and that E is dense in E'^*. However, E'^* is complete since it is a *closed* subspace of the space of *all* the scalar functions on E', with pointwise convergence, which is a complete space. Furthermore E is dense in E'^* for it is sufficient to show that every continuous linear form on E'^* zero on E is identically zero (Section 7, Theorem 3, Corollary 3), but by Proposition 12 every continuous linear form on E'^* comes from an element of E'. The conclusion follows. Notice that if (E, E') is only separated in E, the completion of weak E is still identified with the closure of E in the complete space E'^*, equipped with the topology $\sigma(E'^*, E')$.

EXERCISE 1 Let E be an LCTVS. A sequence (x_i) in E tends weakly to a limit $a \in E$ if and only if a is in the closed convex hull of every subsequence of (x_i).

EXERCISE 2 Let A be a convex subset of an LCTVS. A is weakly compact if and only if every filter base on A formed by closed convex sets has a non-empty intersection.

(Hint: For the sufficiency show first of all that A is weakly precompact since it is weakly bounded; then show that A is weakly complete. For this, if ϕ is a weak Cauchy filter on A, notice that the set of closed convex hulls of $B \in \phi$ is still a weak Cauchy filter base and admits a weak cluster point which is a limit of the filter, and a fortiori of ϕ.)

EXERCISE 3 Let E be a normed TVS of infinite dimension. Show that E is not weakly complete. (There are on E' non-continuous linear forms for the norm topology of E'—see Chapter 1, Section 14, Remarks—and a fortiori non-continuous for $\sigma(E', E)$.)

9 Polarity

DEFINITION 7 *Let (E, E') be a dual system, let A be a subset of E. The polar of A, denoted by A^0 is the set of $x' \in E'$ such that $\mathscr{R}\langle x, x' \rangle \geqslant -1$ for every $x \in A$. There is an analogous definition for the polar of a subset of E'.*

The most inmediate properties of this notion are outlined in

PROPOSITION 14   *Let (E, E') be a dual system.*

1) *The polar of a subset A of E is a subset of E' which is convex, contains the origin, and is weakly closed. It remains unchanged if we replace A by the weakly closed convex hull of $A \cup \{0\}$.*

2) *The transformation $A \mapsto A^0$ is decreasing. We have*

$$(\lambda A)^0 = \frac{1}{\lambda}(A^0)$$

for every real scalar λ. If A and B are two subsets of E, we have

$$(A \cup B)^0 = A^0 \cap B^0.$$

3) *If A is disked, then A^0 is disked and identical to the set of $x' \in E'$ such that $|\langle x, x'\rangle| \leqslant 1$ for every $x \in A$. If A is a cone, A^0 is a cone, identical to the set of $x' \in E'$ such that $\mathscr{R}\langle x, x'\rangle \geqslant 0$ for every $x \in A$. If A is a vector space, A^0 is the vector subspace of E' orthogonal to A, i.e. the set of $x' \in E'$ such that $\langle x, x'\rangle = 0$ for every $x \in A$.*

If A is a subset of E, we call the set of $x' \in E'$ such that $|\langle x, x'\rangle| \leqslant 1$ for every $x \in A$, the *absolute polar* of A, and the set of $x' \in E'$ such that

$$\mathscr{R}\langle x, x'\rangle \geqslant 0$$

for every $x \in A$ the *supplementary cone* of A. From Proposition 4, 3) we know that the polar of a disk is identical to its absolute polar and that the polar of a cone is identical to its supplementary cone. The absolute polar of a set A does not change when A is replaced by its disked hull; from this we see that it is also the polar of the disked hull of A; similarly, the supplementary cone of A is identical to the polar of the cone generated by A.

THEOREM 5 (BIPOLAR THEOREM)   *Let (E, E') be a dual system, A a weakly closed convex subset of E containing the origin. Then $(A^0)^0 = A$.*

This means that if $x_0 \notin A$ there exists an $x' \in E'$ such that

$$\mathscr{R}\langle x, x'\rangle \geqslant -1 \text{ for } x \in A \quad \text{and} \quad \mathscr{R}\langle x_0, x'\rangle < -1.$$

In this assertion, it is only necessary to consider the structure of E as a *real* TVS with the topology weak (Section 1). The existence of a linear form x' as above results immediately from Section 7, Theorem 3. If A is a subset of E, then its polar A^0 and therefore its bipolar $(A^0)^0$ do not change if we replace A by the weakly closed hull of $A \cup \{0\}$; from this we obtain

COROLLARY 1   *Let A be a subset of E. Then the weakly closed convex hull of $A \cup \{0\}$ is identical to the "bipolar" $(A^0)^0$ of A.*

Similarly, the weakly closed disked hull of A is identical to the polar of the absolute polar of A; the weakly closed convex cone generated by A is identical to the supplementary cone of the supplementary cone of A.

The mechanism of polarity as a result of Proposition 14 and Theorem 5, is outlined in the

COMMENT   *Let (E, E') be a dual system. Let $K(E)$ be the set of convex subsets of E which contain the origin and are weakly closed; let $K(E')$ be the analogous set for E'. $K(E)$ and $K(E')$ are ordered by inclusion. If with every $A \in K(E)$ we associate its polar A^0 and with every $A' \in K(E')$ its polar A'^0, we obtain a bijective mapping from $K(E)$ onto $K(E')$. The mapping $A \mapsto A^0$ is an isomorphism from the ordered set $K(E)$ onto the set $K(E')$ with the opposite order: If $A, B \in K(E)$ then $A \subset B$ is equivalent to $A^0 \supset B^0$. The polar of $\overline{\Gamma}(A \cup B)$ is $A^0 \cap B^0$, the polar of $A \cap B$ is $\overline{\Gamma}(A^0 \cup B^0)$. We have $(\lambda A)^0 = \frac{1}{\lambda}(A^0)$ for every real scalar λ.*

Finally, the correspondence outlined above between $K(E)$ and $K(E')$ induces a bijective correspondence between the weakly closed disks of E and E', between the weakly closed convex cones of E and E', and between the weakly closed subspaces of E and E'. These correspondences can also be obtained as stated in Proposition 14, 3).

Notice also the formulae

$$(\overline{\Gamma}(A \cup B))^0 = A^0 \cap B^0 \quad \text{and} \quad (A \cap B)^0 = \overline{\Gamma}(A^0 \cup B^0).$$

The first one is trivial (Proposition 14) and the second an a priori nontrivial result. The second one results from the beginning of the comment (therefore from Theorem 5) and from the first formula applied to A^0 and B^0 instead of A and B. We can also point out that in $K(E)$, $A \cap B$ and $\overline{\Gamma}(A \cup B)$ are respectively the g.l.b. and the l.u.b. of A and B and must therefore be transformed by polarity into the l.u.b., or g.l.b. respectively of A^0 and B^0 by virtue of the beginning of the comment. In cases where A and B are vector spaces V and W, the preceding formulae become

$$\overline{(V + W)^0} = V^0 \cap W^0 \quad \text{and} \quad (V \cap W)^0 = \overline{V^0 + W^0}$$

where the first is a priori trivial but not the second.

Theorem 5 is clearly equivalent to Theorem 3 stated for a vector space with a weak topology. (We can always suppose that the convex set in the statement of Theorem 3 contains 0.) Therefore, the conjunction of Theorems 4 and 5 is exactly equivalent to Theorem 3. Depending on applications, we use either of the parts of Theorem 3, which particular statements are easier to use than Theorem 3 itself. In what follows the bipolar theorem (Theorem 5) will be of special importance.

For the next topics considered, the concept of *absolute polar* will be of more use than that of polar given in Definition 7 (which is useful for example in ordered vector spaces). In all studies involving polarity, the polar of A is what we call here its absolute polar; but if we restrict ourselves to disked sets, which is not hard to do in practice, we have seen that the two notions coincide. The concept of absolute polar, independently introduced by different authors, has been systematically used by Dieudonné and Schwartz. The more general notion of polar (Definition 7) is due to N. Bourbaki.

EXERCISE   Let M be a locally compact space. For every $t \in M$ let $\mathscr{E}_t$ be the linear form $f \mapsto f(t)$ on $C(M)$. Show that the unit ball of the dual $\mathscr{M}(M)$ of $C(M)$ is the weakly closed disked hull of the set of $\mathscr{E}_t$. An element $\mu \in \mathscr{M}(M)$ is said to be *positive* if $\langle f, \mu \rangle \geqslant 0$ for $f \geqslant 0$. Show that the set of $\mu \in \mathscr{M}(M)$ which are positive is the weakly closed convex cone generated by the $\mathscr{E}_t$, and that the set of positive μ such that $\| \mu \| \leqslant 1$ is the weakly closed convex hull of the set of $\mathscr{E}_t$ and 0. From this, conclude that the set of positive μ of norm $= 1$ is the closed convex hull of the $\mathscr{E}_t$. (Show that if $\mu \geqslant 0$, then $\| \mu \| = u(1)$.)

10 The 𝔖-topologies on a dual

PROPOSITION 15   *Let (E, E') be a dual system, A a subset of E. The following conditions are equivalent:*

    a) *A is weakly bounded.*

    b) *For every $x' \in E'$ the set of $\langle x, x' \rangle$, $x \in A$, is bounded.*

    c) *A is weakly precompact.*

The equivalence of a) and b) is a particular case of Chapter 1, Section 7, Proposition 11. On the other hand the equivalence of b) and c) results from the usual characterization of precompact subsets in a uniform space whose uniform structure is the coarsest for which mappings $f_i : E \mapsto E_i$ into uniform spaces, are uniformly continuous (here the f_i are the mappings $x \mapsto \langle x, x' \rangle$ from E into the field of scalars k); we know that the precompact subsets of k are identical to the bounded subsets.

PROPOSITION 16   *Let (E, E') be a dual system, 𝔖 a set of subsets of E. On E' the 𝔖-topology is compatible with the vector structure if and* *only if the $A \in \mathfrak{S}$ are weakly bounded subsets of E. If so, E' equipped with the $\mathfrak{S}$-topology is a locally convex space and a fundamental system of neighborhoods of the origin for this topology is obtained by taking the absolute polars of $A \in \mathfrak{S}$ and the finite intersections of non-zero homothetics of such polars.*

The first assertion is a result of Proposition 15 and Chapter 1, Section 18, Theorem 3; the second assertion follows from the same theorem and the definition of absolute polars. We then see that in particular the $\mathfrak{S}$-topology on E' does not change if we replace the $A \in \mathfrak{S}$ by their weakly closed disked hulls, not even if we add to $\mathfrak{S}$ the weakly closed disked hulls B of finite unions of homothetics of sets $A_i \in \mathfrak{S}$, and finally every weakly closed disk contained in a set such as B. We are then led to a set $\mathfrak{S}_0$ of weakly bounded and weakly closed disks of E, invariant by non-zero homothetics, increasingly directed and containing all the weakly closed disks which are contained in some set $A \in \mathfrak{S}$. We thus obtain a fundamental system of neighborhoods of the origin in E', for the $\mathfrak{S}_0$-convergent topology, by taking the polars of $A \in \mathfrak{S}_0$. But it follows from the bipolar theorem that we cannot add to $\mathfrak{S}_0$ any other weakly closed disks without changing the corresponding topology on E'. In general:

PROPOSITION 17   *Let (E, E') be a dual system, $\mathfrak{S}$ a set of weakly disked and weakly closed subsets of E, invariant by non-zero homothetic, increasingly directed and containing with every set A all the weakly closed disks contained in A. Let τ be a set of weakly closed and weakly bounded disks of E, on E' the $\mathfrak{S}$-topology is finer than the τ-topology if and only if $\tau \subset \mathfrak{S}$.*

For, if $A \in \tau$, then A^0 is a neighborhood of 0 in E' for the τ-topology, therefore (if this topology is coarser than the $\mathfrak{S}$-topology) it contains a set B^0, with $B \in \mathfrak{S}$, from which we conclude that $A \subset B$ and therefore $A \in \mathfrak{S}$.

COROLLARY 1   *Under the conditions of Proposition 17, supposing that τ satisfies the same conditions as $\mathfrak{S}$, on E' the $\mathfrak{S}$-topology and τ-topology are identical if and only if $\mathfrak{S} = \tau$.*

COROLLARY 2   *Let (E, E') be a dual system, let $\mathfrak{S}$ and τ be two sets of bounded subsets of E. On E' the $\mathfrak{S}$-topology is finer than the τ-topology if and only if every $A \in \tau$ is contained in the weakly closed disked hull of the union of a finite number of homothetics of $A_i \in \mathfrak{S}$.*

This corollary is in fact equivalent to Proposition 17, because of the remarks made before its statement.

DEFINITION 8   Let (E, E') be a dual system. The strong topology or the topology of bounded convergence on E' is the topology of uniform convergence on the bounded subsets of E. If E is locally convex, we call the dual space E' with topology of uniform convergence on the bounded subsets of E, the strong dual of E.

Since the bounded weakly closed subsets of a LCTVS E are already bounded for the original topology (Theorem 4'), the strong topology of E' depends only on the dual system (E, E'). According to Definition 8, the strong topology of E' is the finest of the locally convex $\mathfrak{S}$-topologies considered above.

PROPOSITION 17'   Let (E, E') be a dual system. A locally convex topology on E' is a $\mathfrak{S}$-topology for some set $\mathfrak{S}$ of bounded subsets of E if and only if it admits a fundamental system of neighborhoods V_i of 0 disked and closed for $\mathfrak{S}(E', E)$.

The necessity follows from the characterization of neighborhoods of 0 for the $\mathfrak{S}$-topology just given. For the sufficiency we observe that if A_i is the polar of V_i we have $V = A_i^\circ$, therefore the topology considered for E' is the $\mathfrak{S}$-topology, where $\mathfrak{S}$ is the family of A_i. If we start with a LCTVS E we can see by changing the roles of E and E' in the proposition that the topology of E is a $\mathfrak{S}$-topology for some set $\mathfrak{S}$ of subsets of E' (use Theorem 4). We will now consider the $\mathfrak{S}$-topology more closely.

11 The LCTVS as duals having $\mathfrak{S}$-topologies

PROPOSITION 18   Let E be a locally convex space, E' its dual, E^* its algebraic dual. A linear form x' on E is continuous if and only if there exists a disked neighborhood V of 0 in E on which x' is bounded by 1 in absolute value, i.e. such that x' belongs to the polar of V in E^*. A set A' of linear forms on E is equicontinuous if and only if there exists a disked neighborhood V of 0 in E such that all the $x' \in A'$ are bounded by 1 in absolute value on V, i.e. such that A is contained in the polar of V in E^*.

The proof is immediate.

COROLLARY 1   Let E be an LCTVS, A an equicontinuous subset of the dual of E. Then, the weakly closed convex hull of A is still equicontinuous.

COROLLARY 2   Let E be a locally convex space. Then, by polarity, there is *a bijective correspondence between the disked and closed neighborhoods of 0 in E, and the disked equicontinuous weakly closed subsets of E'.*

In fact, the polar of a neighborhood of 0 in E is an equicontinuous subset of E' (Proposition 18), and the polar of an equicontinuous subset of E' is clearly a neighborhood of 0 in E. The corollary follows from the bipolar theorem. Taking into consideration Proposition 16, we find:

COROLLARY 3   *Let E be locally convex, E' its dual. Then the topology of E is the $\mathfrak{S}$-topology where $\mathfrak{S}$ is the set of equicontinuous subsets of E' (or if we wish, the set of weakly closed equicontinuous disked subsets of E').*

In particular, if E is Hausdorff, E appears as the dual of weak E' with a $\mathfrak{S}$-topology, where $\mathfrak{S}$ is as above. The method of topologization studied in Section 10 then gives us *all* the locally convex spaces. The topology of a locally convex space is known once we know the sets of linear forms which are equicontinuous for the topology.

THEOREM 6   *Let E be an LCTVS. Then the equicontinuous subsets of the dual are weakly relative compact subsets.*

According to Proposition 18, we can restrict ourselves to showing that if V is a disked neighborhood of 0 in E, then the polar V^0 of V *in E^** is weakly compact. However, V^0 is a *closed* subset of weak E^* which is a *complete* space (see Proposition 13), therefore V^0 is weakly complete. On the other hand, from Proposition 15, V^0 is weakly precompact, hence the conclusion.

EXERCISE 1   Show the Hahn–Banach theorem II as a consequence of the bipolar theorem (and of the more elementary Theorem 6). (Show first of all that every continuous linear form x' on the subspace F is the restriction of a continuous linear form on E, by applying the bipolar theorem to the disk A consisting of the $x \in F$ such that $| \langle x, x' \rangle | \leqslant 1$: there exists a $y' \in A^0 \subset E'$ not everywhere zero on F with its restriction to F proportional to x', then apply the bipolar theorem and Section 5, Exercise 6, in order to calculate the polar in E' of $V \cap F$ where V is the unit ball associated with the given semi-norm on E).

EXERCISE 2

a) Let E be a *separable* LCTVS. Show that the weakly compact subsets of E' are metrisable.

b) Let E be a metrisable LCTVS; E is separable if and only if the equicontinuous subsets of E' are weakly metrisable (use a) for the necessity and Chapter 1, Section 9, Proposition 16 for the sufficiency). Show that this implies that E' is weakly separable.

c) Let E be normed. The space E is separable if and only if the unit ball of E' is weakly metrisable.

EXERCISE 3   Let E be a vector space, E^* its algebraic dual. There is a bijective correspondence between the semi-norms p on E and the weakly compact disks A of E^* : to p there corresponds the polar of the "unit ball" of p, and conversely to A there corresponds the semi-norm

$$p(x) = \sup_{x' \in A} | \langle x, x' \rangle |.$$

If we start from a TVS E, then for the correspondence indicated above the *continuous* semi-norms are those corresponding to the weakly compact disks contained in E'.

12 Mackey's theorem: general formulation. Bidual of an LCTVS

THEOREM 7 (MACKEY)   *Let (E, E') be a dual system separated in E, let $\mathfrak{S}$ be a set of bounded disked subsets of E, increasingly directed and closed under homothetics. Let E' have a $\mathfrak{S}$-topology. Then the dual of E' is identical to the subspace of the weak completion $\hat{E}$ of E, union of weak closures in $\hat{E}$ of sets $A \in \mathfrak{S}$. A subset of $\hat{E}$ is an equicontinuous set of linear forms on E' if and only if the subset is contained in the weak closure in $\hat{E}$ of a set $A \in \mathfrak{S}$.*

It is sufficient to prove the second assertion. The equicontinuous subsets of the dual of E' are the subsets of the algebraic dual E'^* contained in the polar of a disked neighborhood of 0 (Proposition 18), which we can suppose of the form A^0, $A \in \mathfrak{S}$ (Section 10). By the theorem bipolar, the polar of A^0 in E'^* is identical with the weak closure of the disk A and the conclusion follows.

COROLLARY 1   *Let (E, E') be a dual system separated in E, let $\mathfrak{S}$ be a set of bounded subsets of E and let E' have the $\mathfrak{S}$-topology. Then the dual of E' is the vector subspace of the weak completion $\hat{E}$ of E generated by the weakly closed disked hulls in $\hat{E}$ of the $A \in \mathfrak{S}$.*

We can suppose the $A \in \mathfrak{S}$ disked; it is sufficient to show that if we have a finite set of elements A_i of $\mathfrak{S}$ and of scalars λ_i, then the weakly closed disked hull of $\cup \lambda_i A_i$ in $\hat{E}$ is contained in the hull considered in the corollary. But, if $\bar{A}_i$ is the weak closure of A_i in $\hat{E}$, it is a weakly compact disk in $\hat{E}$, therefore $\Sigma \lambda_i \bar{A}_i$ is a weakly compact disk in $\hat{E}$. It contains $\cup \lambda_i A_i$ and therefore it contains the weakly closed disked hull of this set in $\hat{E}$, a weak hull which is contained in the vector space generated by the $\bar{A}_i$.

COROLLARY 2   *Let (E, E') be a dual system separated in E, let $\mathfrak{S}$ be a set of bounded subsets of E and let E' have the $\mathfrak{S}$-topology. Then:*

1) *the dual of E' is contained in E if and only if the weakly closed disked hulls (in E) of the $A \in \mathfrak{S}$ are weakly compact.*

2) *the dual of E' contains E if and only if the vector space generated by the weakly closed disked hulls (in E) of the $A \in \mathfrak{S}$ is identical to E.*

3) *therefore, the dual of E' is identical with E if and only if the two preceding conditions are verified.*

The case of the strong topology on E' is particularly important.

DEFINITION 9   *Let E be an LCTVS. The bidual of E denoted by E'' will be the dual of the strong dual E' (see Section 10, Definition 8), with the topology of uniform convergence on the equicontinuous subsets of E'.*

We verify immediately that the equicontinuous subsets of E' are strongly bounded, therefore the topology on E'' is locally convex. Since the strong dual of E depends only on the dual system (E, E'), it follows that the bidual is known, except for its topology, once the dual system (E, E') is known. Thus, if we start from a dual system (E, E') we can still call the strong dual of E' the bidual of E, (without an explicit topology, a priori). If E is a Hausdorff LCTVS, it is identified with a vector subspace of E'', the topology on E being that induced by E''; and E'' is in turn identified with a vector subspace of the weak completion $\hat{E}$ of E (but, clearly, the topology of E'' will not be, in general, that induced by $\hat{E}$). In short, from Mackey's theorem, E'' is the union of weak closures in $\hat{E}$ of bounded subsets of E, and more generally the equicontinuous subsets of E'' (considered as a strong topological dual of E') are the subsets of E'' contained in the weak closure of a bounded subset of E.

DEFINITION 10   *Let E be an LCTVS. E is said to be reflexive if it is Hausdorff and identical with its bidual.*

Similarly, if (E, E') is a dual system we can call it reflexive in E if E-weak is reflexive (the reflexivity of an LCTVS depends on the duality). The past part of corollary 2 of Theorem 7 gives

THEOREM 8   *Let E be an LCTVS. The space E is reflexive if and only if its bounded subsets are weakly relatively compact.*

COROLLARY   *Let E be a reflexive LCTVS, F a closed vector space, then F is reflexive.*

In fact, since every continuous linear form on F is the restriction of a continuous linear form on E, the weak topology of F is induced by the weak topology of E. On the other hand, F is weakly closed (Section 8, Theorem 4); the corollary results from Theorem 8.

13 Topologies compatible with a duality. The Mackey topology

In this Section, we simply state in a different language the last part of Corollary 2 of Theorem 7 (Mackey's theorem).

DEFINITION 11   *Let (E, E') be a dual system separated in E'. A topology on E is said to be compatible with the duality (E, E') if it is locally convex and if the dual of E for this topology is identical with E'.*

Thus, the weak topology $\sigma(E, E')$ is compatible with the duality (Section 8, Proposition 12) and is clearly the coarsest on E compatible with the duality.

THEOREM 9   *Let (E, E') be a dual system separated in E'. A topology on E is compatible with the duality (E, E') if and only if it is a $\mathfrak{S}$-topology, where $\mathfrak{S}$ is a set of weakly compact disks in E' covering E'.*

The condition is necessary by Proposition 18, Corollary 3; it is sufficient by Theorem 7 (Mackey's theorem), Corollary 3 (where we only reverse the roles of E and E'). Therefore, among the topologies on E compatible with a duality, there is a finest one which corresponds to $\mathfrak{S}$, the set of *all* the weakly compact disks in E'.

DEFINITION 12   *Let (E, E') be a dual system, separated in E; we call the topology of uniform convergence on the weakly compact disks of E the Mackey topology on E, and denote it by $\tau(E, E')$. We call this topology $\tau(E, E')$, the Mackey topology associated with the topology of a given LCTVS E (where E' is the dual of E).*

Theorem 9 can then be stated as follows:

COROLLARY   *The topologies on E compatible with the duality (E, E') are exactly the locally convex topologies included between the weak topology $\sigma(E, E')$ and the Mackey topology $\tau(E, E')$.*

Proof   For such a topology, the dual of E' is included between the duals for the weak topology and the Mackey topology, i.e. it is identical with E'.

In the sequel we will refer to any of Theorems 7, 8 or 9 as "Mackey's theorems". We will see in Chapter 3 that the topology of a metrisable LCTVS always coincides with Mackey's topology.

PROPOSITION 19   Let E be an LCTVS. A disked and absorbing subset V of E is a neighborhood of the origin for the Mackey topology $\tau(E, E')$ if and only if every linear form on E, bounded by 1 in absolute value on V, is continuous.

The necessity results from Mackey's theorem, since the considered linear form is continuous for $\tau(E, E')$. Conversely, if the polar of V in E^* is contained in E', we will show that V is a neighborhood of 0 for $\tau(E, E')$. We can suppose that V is the unit ball associated with some semi-norm on E (Section 4, Theorem 1), therefore (by the bipolar theorem applied to V in the space E with the preceding semi-norm) V is the polar in E of V^0. Since V^0 is a weakly compact disk in E' (apply Theorem 6 of Section 11 to E considered as a semi-normed space), it follows that V is a neighborhood of 0 for $\tau(E, E')$.

EXERCISE 1   Show that in the statement of Theorem 9 and in Definition 12, we can replace the word "disk" by "convex set" (see Section 5, Exercise 4).

EXERCISE 2   Let E be a complete Hausdorff LCTVS, (x_i) a bounded sequence in E, u the continuous linear mapping of l^1 into E defined as in Chapter 1, Section 9, Exercise 6. Show that u is continuous for the topology $\sigma(l^1, c_0)$ (see Chapter 1, Section 9, Exercise 7) and for the weak topology of E if and only if (x_i) tends weakly towards 0. From this conclude that if (x_i) is a sequence tending weakly to 0 in a complete Hausdorff LCTVS, then its closed convex hull (or its closed disked hull) is the set of sums $\Sigma \lambda_i x_i$ where (λ_i) runs through the set of positive sequences such that $\Sigma \lambda_i = 1$ (or where (λ_i) runs through the unit ball of l^1). (Show that the two preceding sets are weakly compact, therefore closed, applying Theorem 6 to c_0 and its dual l^1.)

EXERCISE 3
a) Let E be a complete Hausdorff LCTVS, F a vector subspace, x a point in the closure of F. Show that we can find a bounded sequence (x_i) in F, and a positive sequence of summable scalars (λ_i) such that we have $x = \Sigma \lambda_i x_i$ (this series converges, see Chapter 1, Section 9, Exercise 6). Show that if (μ_i) is a sequence of scalars > 0 we can assume above that $\lambda_i \leqslant \mu_i$ for every i.

b) From this conclude that if x is a point in $\bar{F}$, we have $x \in \bar{\Gamma}(K)$,

F

where K is some compact subset of F (take for K a sequence tending to 0 in F and containing the origin).

c) Conclude from b) that if E is a metrisable LCTVS, E is complete if and only if the closed convex hull of every compact subset of E is compact (for the necessity see Section 5, Exercise 5).

d) Conclude from a) that if E is a normed space E is complete if and only if every absolutely convergent series in E is convergent.

EXERCISE 4

a) Let E be a complete Hausdorff LCTVS, or more generally an LCTVS whose closed and bounded subsets are complete. We equip E' with the topology of compact convergence. Show that the dual of E' is identical with E (use Mackey's theorem and Section 5, Exercise 5).

b) Let E be a metrisable LCTVS, equip its dual E' with the topology of compact convergence. Show that the dual of E' can be identified with the completion of E (see Exercise 3), and that it is identical with E if and only if E is complete. From this, conclude that Theorem 9 does not hold if the word "disked" is removed from the statement.

EXERCISE 5   Let E be an LCTVS whose topology is identical with the associated Mackey topology. Show that this is also true for the completion of E.

EXERCISE 6   Let μ be a bounded measure on a locally compact space M, let A be a convex subset of $L^\infty = L^\infty(\mu)$. Show that if f is in the closure of A for the weak topology of L^∞ considered as dual of L^1, there exists a sequence (f_i) in A such that for every p, $1 \leq p < +\infty$, f_i tends to f in the sense of the topology induced by L^p. (Notice that the topology induced by L^p is the topology of uniform convergence on the unit ball of $L^{p'}$, which is a *weakly compact subset* of L^1, $L^{p'}$ being reflexive. (We may assume $p > 1$); therefore, the topology induced by L^p is coarser than the Mackey topology $\tau(L^1, L^\infty)$, so that it is sufficient to apply Mackey's theorem and Theorem 4 of Section 9.) Show that, conversely, if A is bounded in the normed space L^∞, then the closure of A in the normed space L^1 is identical with its weak closure in L^∞ (apply Theorem 6 to A).

EXERCISE 7   Let M be an LCTVS, let $E = C_0(M)$; then E' is the space of bounded measures on M. We identify $l^1(M)$ with the space of discrete bounded measures on M, and M with a subset of $l^1(M)$ associating with every $x \in M$ the measure of mass $+ 1$ at the point x. For every $X \in E''$ let f_X be the restriction of X to M; it is a bounded function on M, i.e.

an element $f_X \in l^\infty(M)$. The restriction of X to $l^1(M)$ is exactly f_X when we identify $l^\infty(M)$ with the dual of $l^1(M)$ (Chapter 1, Section 9, Exercise 7).

1) Show that every bounded function on M, $f \in l^\infty(M)$, is of the form f_X, where X is an element of E'' of norm $\| f \|$. (Identify f with a linear form on the subspace $l^1(M)$ of E' and apply the Hahn–Banach theorem; or, proceed directly by proving that the unit ball of $E = C_0(M)$ is dense for the topology of pointwise convergence in the unit ball of $l^\infty(M)$.) From this, conclude that if M is infinite, i.e. E of infinite dimension, then E is not reflexive.

2) A space $L^1(\mu)$ (constructed on a positive measure μ on a locally compact space) is not reflexive, unless it is of finite dimension, i.e. μ is the sum of a finite number of point masses. (Notice that the dual of L^1 which is L^∞, is isomorphic to a space $C(M)$, M compact, by virtue of a classical theorem of Stone-Gelfand, and also, if $L^1(\mu)$ is reflexive, its dual is, whence the conclusion using 1).)

14 The completion of an LCTVS

THEOREM 10   *Let (E, E') be a separated dual system, $\mathfrak{S}$ an increasingly directed set of bounded disks of E such that the vector space generated by their union is identical with E. Equip E' with a $\mathfrak{S}$-topology and let $\hat{E}'$ be the set of linear forms on E whose restrictions to all the $A \in \mathfrak{S}$ are weakly continuous. Then $\hat{E}'$ is a complete Hausdorff LCTVS and E' is a dense topological vector subspace of $\hat{E}'$; in other words, the TVS $\hat{E}'$ is the completion of E' (for the $\mathfrak{S}$-topology).*

We first point out the following corollaries:

COROLLARY 1   *Under the conditions of Theorem 10, E' equipped with a $\mathfrak{S}$-topology is complete if and only if every linear form on E, whose restrictions to $A \in \mathfrak{S}$ are weakly continuous, are weakly continuous.*

COROLLARY 2   *Let E be a Hausdorff LCTVS; then the completion of E can be identified with the space of linear forms on E' whose restrictions to the equicontinuous subsets are weakly continuous, when this space has the uniform topology of convergence on the equicontinuous subsets of E'.*

COROLLARY 3   *Let E be a Hausdorff LCTVS. The space E is complete if and only if every linear form on E', whose restrictions to the equicontinuous subsets are weakly continuous, is already weakly continuous, i.e. is a member of E.*

(The last statement is the most useful.)

Proof

1) We consider first of all the particular case where the $A \in \mathfrak{S}$ are weakly compact (this is in practice the most important case since it contains Corollary 3). For every $x' \in \hat{E}'$ and $A \in \mathfrak{S}$, $x'(A)$ is a compact set and is therefore bounded in the field of scalars, therefore $\hat{E}'$ is locally convex for the $\mathfrak{S}$-topology. Also, $\hat{E}'$ is complete since it is a *closed* uniform subspace of the space of all mappings of E into the field K equipped with the $\mathfrak{S}$-topology, which is a complete space. The topology of E' is clearly that induced by $\hat{E}'$. To prove that E' is dense in $\hat{E}'$ we must verify that every continuous linear form on $\hat{E}'$ vanishing on E' is identically zero (Section 7, Theorem 3, Corollary 3). Now, the $A \in \mathfrak{S}$ are still compact for $\sigma(E, \hat{E}')$ (since the $x' \in \hat{E}'$ have restrictions to $A \in \mathfrak{S}$ which are already continuous for $\sigma(E, E')$) so the dual of $\hat{E}'$ is identical with E (Mackey's theorem), and the conclusion follows.

2) General case—we use the following lemma, interesting in itself.

LEMMA　*Let E be an LCTVS, A a convex symmetric subset of E, u a linear mapping of E into an LCTVS F. For the restriction of u to A to be uniformly continuous it is sufficient that it be continuous at the origin.*

In fact, we must find for every neighborhood V of 0 in F a neighborhood U of 0 in E such that $x, y \in A$, $x - y \in U$ imply

$$u(x) - u(y) \in V \quad \text{that is} \quad u(x - y) \in V.$$

We will have $x - y \in A - A = 2A$, therefore such a U exists if $u(2A \cap U) \subset V$, that is

$$u(A \cap \tfrac{1}{2}U) \subset \tfrac{1}{2}V,$$

which is true.

Applying the lemma to Theorem 10, we see that for every $x' \in \hat{E}'$ the restriction of x' to every $A \in \mathfrak{S}$ is uniformly weakly continuous, therefore it can be extended by uniform continuity to a function x'_A on the weak closure $\bar{A}$ of A in E'^*. Let E_1 be the vector space generated by the $\bar{A}$, let $\mathfrak{S}_1$ be the set of weakly compact disks $\bar{A}$ in E_1. If $x \in E_1$, we have $\lambda x \in \bar{A}$ for some $\lambda > 0$ and $A \in \mathfrak{S}$, and we verify immediately that the expression

$$\frac{1}{\lambda}x'_A(\lambda x)$$

does not depend on the choice of the pair (λ, A) (since $\mathfrak{S}$ is directed); let $\langle x, x' \rangle$ be its value and notice that the function $\langle x, x' \rangle$ thus defined on $E_1 \times E'$ is bilinear. Finally, for the described pairing of E_1 and $\hat{E}'$,

$\hat{E}'$ can be identified with the space of linear forms on E_1 whose restrictions to the $A_1 \in \mathfrak{S}_1$ are continuous, and the old $\mathfrak{S}$-topology is identical with the $\mathfrak{S}_1$-topology. Since the system (E_1, E'), $\mathfrak{S}_1$, $\hat{E}'$ satisfies the conditions of the first part of the proof, it follows that $\hat{E}'$ is locally convex and is the completion of E'.

EXERCISE 1   Let E be an LCTVS, A a disked subset of E, $x' \in E^*$.

a) the restriction of x' to A is continuous if and only if, for every $\varepsilon > 0$, the intersection of A with the hyperplane of equation $\langle x, x' \rangle = \varepsilon$ is relatively closed in A. (Use the Lemma and show that if U is a disked neighborhood of 0 in E such that $U \cap A$ does not meet V, then $|\langle x, x' \rangle| < \varepsilon$ for $x \in U \cap A$; notice that we only use the fact that $0 \notin \overline{A \cap V_\varepsilon}$).

b) Conclude from a) that the restriction of x' to A is continuous if and only if it is continuous for the topology induced by the weak topology of E.

EXERCISE 2   Let E be a Hausdorff LCTVS, A a convex subset of E. Show that the completion of A can be identified with the set of $x \in E'^*$ which are in the weak closure of A and which define linear forms on E' whose restrictions to the equicontinuous subsets of E' are weakly continuous. (Use Theorem 10, Corollary 2 and the corollary of Theorem 4 applied to the completion of E.)

EXERCISE 3   Deduce Proposition 13 from Theorem 10.

EXERCISE 4   Let E be a complete Hausdorff LCTVS. If E' equipped with the strong topology is reflexive, the E is reflexive (show that every strongly continuous linear form on E' is weakly continuous, by using Theorem 10, Corollary 3). Show that it is sufficient to suppose that the closed and bounded subsets of E are complete (proceed as before, but use Theorem 7 and Exercise 2).

REMARK   These results are also particular cases of more general results, to be studied in Section 18.

EXERCISE 5   Let E be a vector space, A a convex symmetric subset. Show that if two locally convex topologies on E induce on A the same system of neighborhoods of 0, they also induce the same uniform structures. (This is a corollary of the lemma in this Section.)

15 Duality for subspaces, quotients, products, projective limits

Let E be an LCTVS, F a vector subspace. If to every $x' \in E'$ we assign its restriction to F, we obtain a linear mapping from E' into F' whose kernel is the orthogonal F^0 of F in E'; this mapping is furthermore from E' onto F' (Hahn–Banach theorem II, Corollary 1), therefore F' can be identified with E'/F^0; then the equicontinuous subsets of F' are the canonical images of equicontinuous subsets of E' (Hahn–Banach II, Corollary 1). From the identity $F' = E'/F^0$ it follows trivially that the weak topology $\sigma(F, F')$ of F is identical with the topology induced by the weak topology $\sigma(E, E')$ of E. On the other hand, it is trivial that the dual of E/F can be identified with the space of continuous linear forms on E vanishing on F, that is, to F^0; the equicontinuous subsets of the dual of E/F can then be identified with the equicontinuous subsets of E' contained in F^0. If E_s stands for E with the $\sigma(E, E')$ topology and if we apply the last assertion to E_s/F, we find that the equicontinuous subsets of the dual of E_s/F are the subsets of F^0 which are contained in and are bounded in a finite dimensional vector space. That is, they are exactly the equicontinuous subsets of the dual of $(E/F)_s$; consequently, the weak topology associated with the LCTVS E/F is identical with the quotient topology of the weak topology of E. Summing up:

PROPOSITION 20   *Let E be an LCTVS, F a vector subspace. Let F have the topology induced by E, and E/F the quotient topology. Then the dual of F can be identified with E'/F^0, the equicontinuous subsets of this dual being the canonical images of equicontinuous subsets of E', and the weak topology of F being identical with the topology induced by the weak topology of E. The dual of E/F can be identified with F^0, the equicontinuous subsets of this dual being the equicontinuous subsets of E' contained in F^0. Finally, the weak topology of E/F is identical with the quotient topology of the weak topology of E.*

The results concerning the topologies on the duals are summed up in

PROPOSITION 21   *Let E be an LCTVS, F a vector subspace, $\mathfrak{S}$ a set of bounded subsets of E, $\mathfrak{S}_1$ the set of intersections $A \cap F$ where A runs through $\mathfrak{S}$; let $\phi(\mathfrak{S})$ be the set of canonical images of the $A \in \mathfrak{S}$ in E/F.*

1) The $\phi\mathfrak{S}$-topology in the dual F^0 of E/F, is identical to the topology induced by the $\mathfrak{S}$-topology in E'.

2) Suppose E Hausdorff, F closed, $\mathfrak{S}$ increasingly directed and the $A \in \mathfrak{S}$ disked and weakly compact. Then in the dual E'/F^0 of F, the

$\mathfrak{S}_1$-*topology is identical to the quotient topology of the* $\mathfrak{S}$-*topology in* E'.

1) is trivial. In order to show 2) we may obviously suppose $\mathfrak{S}$ closed under homothetics. Consider on E'/F^0 the quotient $\mathfrak{S}$-topology, whose dual is the orthogonal of F^0 in the dual of E' by Proposition 21; however, the dual of E' is the vector space $\cup A,\ \in \mathfrak{S}$ (Mackey's theorem) and the orthogonal of F^0 in this space is the trace of F on this space (F is closed); furthermore, the equicontinuous subsets of the dual of E'/F^0 are the subsets of the dual which are equicontinuous in the dual of E' (Proposition 20), i.e. contained in an $A \in \mathfrak{S}$. Those are also the sets contained in an $A_1 \in \mathfrak{S}_1$, therefore exactly the sets which are equicontinuous when E'/F^0 is given $\mathfrak{S}_1$-convergence; from this follows the equality of the two topologies considered for E'/F^0.

COROLLARY 1 *Let* E *be an* LCTVS, F *a vector subspace. Then the weak topology of the dual of* E/F *is identical to the topology induced on* F^0 *by the weak topology of* E'. *If* F *is closed, the weak topology of* $F' = E'/F^0$ *is identical to the quotient topology of the weak topology of* E'.

COROLLARY 2 *Let* E *be a reflexive* LCTVS, F *a closed vector subspace. Then* F *is reflexive and the strong dual of* F *can be identified with the quotient of the strong dual* E' *of* E *by the subspace* F^0.

This is shown by jointly applying Proposition 21 and the Corollary of Theorem 8.

We remark that in the general case where the $A \in \mathfrak{S}$ are not supposed to be weakly compact, the $\mathfrak{S}_1$-topology on the dual of F is coarser than the quotient $\mathfrak{S}$-topology in E' and can be strictly coarser; thus, we can find a metrisable and complete space E and a closed vector subspace F such that the strong dual of F is not identified with a quotient of the strong dual E'. We point out also that by Proposition 21 and Section 1, Proposition 17, Corollary 2, the topology of the strong dual of E/F, a priori finer than the topology induced by strong E', is identical to the latter if and only if every bounded subset of E/F is contained in the closure of the canonical image of a bounded subset of E; this condition is not necessarily verified even when E is a reflexive metrisable and complete space. However, we will see in Section 17, Proposition 32, that the strong topologies are well behaved when E is a *normed* space.

Consider a finite family (E_i) of LCTVS; a linear form x' on the product is bijectively determined by the system (x_i') of restriction to the subspaces E_i since we will have

$$\langle(x_i), x'\rangle = \sum_i \langle x_i, x'_i\rangle.$$

It is clear that x' is continuous if and only if the x'_i are continuous linear forms on the E_i and x' runs through an equicontinuous set if and only if each x'_i does. Therefore,

PROPOSITION 22   *Let (E_i) be a finite family of LCTVS; then the dual of ΠE_i can be identified by the pairing described above, with the product $\Pi E'_i$ of duals. The equicontinuous subsets of the dual of ΠE_i are those contained in a product ΠA_i where for every i, A_i is an equicontinuous subset of E'_i.*

COROLLARY 1   *Let (E_i) be a family of LCTVS; then the dual of the product ΠE_i can be identified with the direct sum $\Sigma E'_i$ of the duals. The equicontinuous subsets of the dual of ΠE_i are those contained in the sum of a finite number of equicontinuous subsets of E'_i.*

We have in fact an obvious linear mapping of $\Sigma E'_i$ into the dual of $E = \Pi E'_i$ which is trivially bijective. A continuous linear form on ΠE_i is bounded above by 1 in absolute value over a set V defined by the conditions: $x_i \in V_i$ for every $i \in J$ where J is a finite subset of I and the V_i neighborhoods of 0 in the E_i. It follows that x' vanishes on the vector subspace $\Pi_{i \notin J} E_i$ of E, and therefore comes from a continuous linear form on the quotient space which can be identified with $\Pi_{i \in J} E_i$. Proposition 22 gives the desired result. We proceed likewise for the equicontinuous sets of linear forms.

COROLLARY 2   *Let (E_i) be a family of LCTVS; then the weak topology of ΠE_i is the product of the weak topologies of the E_i.*

Taking into consideration Theorem 8 (Mackey's theorem), we thus obtain:

COROLLARY 3   *Under the conditions of Corollary 2, ΠE_i is reflexive if and only if the E_i are reflexive.*

COROLLARY 4   *Let (E_i) be a finite family of LCTVS, E their product; for every i, let $\mathfrak{S}_i$ be a set of bounded subsets of E_i and $\mathfrak{S}$ the set of subsets $\Pi_i A_i$ of E with $A_i \in \mathfrak{S}_i$ for every i. Then the dual of E equipped with the $\mathfrak{S}$-topology can be identified with the product of the duals of the E_i equipped with the $\mathfrak{S}_i$-topology.*

This corollary implies in particular that the weak dual of $\Pi\, E_i$ is the product of the weak duals of E_i and that the strong dual of $\Pi\, E_i$ is the product of the strong duals of the E_i (i finite).

The case of the infinite product is interesting in order to prove.

THEOREM 11   *Let E be a vector space, (E_i) a family of* LCTVS, *and for every i, u_i a linear mapping of E into E_i. We assign to E the coarsest topology for which the u_i are continuous. Then the dual of E is identical to the set of linear forms of the form $\sum\limits_{i \in J} x_i' \circ u_i$ where J is a finite subset of the set of indices, and for every $i \in J$, x_i' is an element of the dual of E_i. We obtain the equicontinuous subsets of the dual of E by taking J fixed and letting x_i' run through an equicontinuous subset of the dual of E_i (for every $i \in J$).*

The proof reduces to the case where E is Hausdorff and can therefore be identified with a topological vector subspace of $\Pi\, E_i$. Then the equicontinuous sets of linear forms on E are the restrictions to E of the equicontinuous sets of linear forms on $\Pi\, E_i$ (Proposition 20), which are characterized by Corollary 1 of Proposition 22. Theorem 11 follows.

The following is a useful particular case of Theorem 11. Let E and F be LCTVS, L a vector space of linear mappings of E into F, L_s is the space L with the topology of pointwise convergence, i.e. the coarsest for which the mappings $u \rightarrow u(x)$ from L into F (where $x \in E$) are continuous. Theorem 11 tells us that the continuous linear forms on L_s are the forms of type

$$u \mapsto \sum \langle u(x_i), y_i' \rangle,$$

where (x_i) is a finite sequence in E, (y_i') a finite sequence in F'. Interpreting the $u \in L$ as linear forms on $E \times F'$, i.e. elements of the algebraic dual of $E' \otimes F$, and introducing

$$v = \sum x_i \otimes y_i' \in E \otimes F',$$

we see that the continuous linear forms on L_s are the forms of type $u \mapsto \langle u, v \rangle$, where $v \in E \otimes F'$. That is, $E \otimes F'$ can be canonically mapped *onto* the dual of L_s; this mapping is bijective provided L contains $E' \otimes F$, and E is Hausdorff, as we can easily verify. Thus:

PROPOSITION 23   *Let E, F be* LCTVS, *E Hausdorff, let L be a vector space of linear mappings of E into F containing $E' \otimes F$. Then the dual of L for the topology of pointwise convergence can be identified with $E \otimes F'$.*

COROLLARY   *The dual of L_s does not change when the topology of F is replaced by another with the same dual. For example if it is replaced by the weak topology associated with the initial topology of F.*

It then follows for example that for a convex subset of L, the closure for the topology of pointwise convergence is identical to the closure for the topology of pointwise weak convergence (i.e. the topology of pointwise convergence when we impose a weak topology on F).

EXERCISE 1

a) Let E be a locally convex, metrisable and complete space, F a closed vector subspace. Show that on the dual of E/F, the topology of compact convergence is identical to the topology induced on F^0 by the topology of compact convergence in E' (use Chapter 1, Section 14, Exercise 2).

b) Let E be an LCTVS whose bounded closed subsets are complete, F a closed vector subspace. Show that on the dual of F, the topology of compact convergence is identical to the quotient topology of the topology of compact convergence in E' modulo F^0 (use Section 5, Exercise 5).

EXERCISE 2   Let E be a complete Hausdorff LCTVS whose topology is defined as the coarsest for which linear mappings u_i from E into *reflexive* spaces E_i are continuous. Show that E is reflexive (use Proposition 21, Corollary 2 and Proposition 22, Corollary 3). Show that the result remains valid when we only suppose that the bounded closed subsets of E are complete.

16 The transpose of a linear mapping; characterization of homomorphisms

PROPOSITION 24   *Let E and F be two LCTVS, u a linear mapping from E into F. The mapping u is continuous for the weak topologies if and only if for every $y' \in F'$ the form $y' \circ u$ on E is continuous.*

The proof is trivial.

COROLLARY 1   *Let u be a linear mapping from an LCTVS E into another one F. If u is continuous, it is also continuous for the weak topologies.*

The dual systems (E, E') and (F, F') are the really important elements of Proposition 24. The condition given in it for u to be weakly continuous means also that the algebraic transpose of u (which is a linear mapping u^* from the algebraic dual F^* of F into the algebraic dual E^* of E defined by the classical formula

$$u^*y' = y' \circ u,$$

or, explicitly,

$$\langle x, u^*y' \rangle = \langle ux, y' \rangle)$$

maps the dual F' of F into the dual E' of E. From now on, we will call the linear mapping u' of F' into E', induced by the algebraic transpose of u, the *transpose* of a weakly continuous linear mapping u of E into F.

COROLLARY 2   *Let (E, E') and (F, F') be dual systems separated in E' and F', u a weakly continuous linear mapping from E into F; then the transpose u' of u is a weakly continuous linear mapping from F' into E', whose transpose is u, when E and F are Hausdorff (and when E', F' are equipped with their weak topologies).*

It follows that in the case of separated dual systems (E, E'), (F, F') the mapping $u \longmapsto u'$ defines an isomorphism from the vector space $L(E, F)$ of weakly continuous linear mappings from E into F, onto the vector space $L(F', E')$ of weakly continuous linear mappings from F' into E'; we shall see that essentially what is true for elements, subsets, etc., of $L(E, F)$ can be interpreted, simply, by transposition in terms of elements, subsets, etc., of $L(F', E')$.

Recall that if we consider the transpose of the composition of a sequence of weakly continuous operators:

$$E \to F \to \ldots \to G,$$

we could equally well have considered the composition of the sequence of transposed mappings:

$$G' \to \ldots \to F' \to E'.$$

In particular, the transposition defines an isomorphism of the *algebra* $L(E, E)$ of weakly continuous endomorphisms from E onto the contravariant algebra of $L(E', E')$ of continuous endomorphisms of E':

$$(uv)' = v' \circ u'.$$

Let M be a set of linear mappings from a vector space E onto a vector space F, A a subset of E, B a subset of F; we set

$$M(A) = \bigcup_{u \in M} u(A) \qquad M^{-1}(B) = \bigcup_{u \in M} u^{-1}(B).$$

We then have the following trivial proposition:

PROPOSITION 25   *Let (E, E'), (F, F') be dual systems separated in E' and F', u a weakly continuous linear mapping from E into F, A a subset of F'. Then we have*

$$(u(A))^0 = u^{1-1}(A^0) \qquad u^{-1}(B^0) = (u'(B))^0.$$

More generally, if M is a set of weakly continuous linear mappings from E into F, M' the set of transposed mappings, we get

$$(M(A))^0 = M'^{-1}(A^0), \qquad M^{-1}(B^0) = (M'(B))^0$$

(the superscript ^{0} stands for absolute polar).

COROLLARY   *Under the conditions of Proposition 25, the weakly closed disked hull of $u(A)$ (resp. of $M(A)$) is identical to the polar of $u'^{-1}(A^0)$ (resp. of $M'^{-1}(A^0)$).*

This follows, in fact, from Proposition 25 and the bipolar theorem. In particular, choosing $A = E$, we find:

PROPOSITION 26   *Let (E, E') and (F, F') be dual systems separated in E' and F', let u be a weakly continuous mapping from E into F. Then the weak closure of $u(E)$ is the orthogonal of the kernel of u'; if E and F are separated, the kernel of u is the orthogonal of $u'(F')$.*

The second assertion can be obtained by exchanging the roles of E and E' and of F and F'. In particular:

COROLLARY   *Let (E, E') and (F, F') be separated dual systems, u a weakly continuous linear mapping from E into F. For $u(E)$ to be weakly dense in F, it is necessary and sufficient that u' be one-to-one; u is one-to-one if and only if $u'(F')$ is weakly dense in E'.*

We cannot use this corollary to interpret by transposition the situation $u(E) = F$; such an interpretation is included in

PROPOSITION 27   *Let (E, E') and (F, F') be dual systems separated in E' and F', let u be a weakly continuous linear mapping from E into F. For u to be a weak homomorphism it is necessary and sufficient that $u'(F')$ be a weakly closed subspace of E'.*

Proof   Let N be the kernel of u, v the mapping of E/N into F obtained from u. Since the weak dual of E/N can be identified with the closed topological vector subspace N^0 of weak E', and u' is obtained by composing v' with the identity mapping from N^0 into E', the hypothesis that $u'(F')$ be weakly closed is equivalent to the same hypothesis on v; on the other hand, since the quotient topology of E weak by N is identical to the corresponding weak topology on E/N (Proposition 20), to say that u is a weak homomorphism is to say that v is a weak isomorphism. We thus prove the proposition for the *bijection v* instead of u. But in this case the proposition is obvious (if u is a weak isomorphism, then u' maps F' onto E', by Proposition 20, i.e. the Hahn–Banach theorem; the converse is trivial by the definition of weak topologies).

COROLLARY   *Let (E, E'), (F, F') be separated dual systems, u a weakly continuous linear mapping from E into F. For u to be a weak isomorphism,* *it is necessary and sufficient that*

$$u'(F') = E';$$

for $u(E) = F$ it is necessary and sufficient that u' be a weak isomorphism.

So far we have used dual systems and weak topologies. Now we introduce other locally convex topologies on the vector spaces.

PROPOSITION 28   *Let (E, E') and (F, F') be dual systems separated in E', F'. We equip E (resp. F) with the topology of uniform convergence on a set $\mathfrak{S}$ (resp. $\mathfrak{S}$) of weakly bounded subsets of E' (resp. F'). We suppose that every set contained in a homothetic of the weakly closed disked hull of a finite number of elements of $\mathfrak{S}$, belongs to $\mathfrak{S}$. Let u be a weakly continuous linear mapping from E into F. For u to be continuous it is necessary and sufficient that $u'(\mathfrak{T}) \subset \mathfrak{S}$, i.e. that for every $B \in \mathfrak{T}$ we have $u'(B) \in \mathfrak{S}$. More generally, a set M of weakly continuous linear mappings from E into F is equicontinuous if and only if for every $B \in \mathfrak{T}$ we have $M'(B) \in \mathfrak{S}$ (M' is the set of transposes of $u \in M$).*

Proof   To say that M is equicontinuous is to say that for every neighborhood V of 0 in F, $M^{-1}(V)$ is a neighborhood of 0 in E. We can evidently suppose that V is of the form B^0, where $B \in \mathfrak{T}$. By the last formula of Proposition 25, this means that every $(M'(B))^0$ is a neighborhood of 0 in E, therefore (Section 8) that $M'(B) \in \mathfrak{T}$, hence the conclusion.

COROLLARY 1   *Let E and F be LCTVS, u a linear mapping from E into F. For u to be continuous it is necessary and sufficient that for every equicontinuous subset B of F', the set of $y' \circ u$ where y' runs through B, be an equicontinuous set of linear forms on E (i.e. that u be weakly continuous and u' transforms the equicontinuous subsets of F' into equicontinuous subsets of E').*

COROLLARY 2   *Let E and F be LCTVS, E being equipped with the Mackey topology $\tau(E, E')$. Then the linear mappings from E into F that are continuous, are identical with those that are weakly continuous.*

For, if u is continuous, it is weakly continuous (Proposition 24, Corollary 1) and, conversely, if u is continuous, its transpose is weakly continuous, therefore it transforms the weakly compact disks (and, a fortiori, the weakly closed disked equicontinuous subsets of F') into weakly compact disks of E' and therefore into equicontinuous subsets of E'.

COROLLARY 3   *Let u be a continuous linear mapping from a Hausdorff*

LCTVS E into another, F. Then its transpose u' is continuous when we equip E' and F' with:

a) *the strong topologies;*

b) *the Mackey topologies* $\tau(E', E)$ *and* $\tau(F', F)$*;*

c) *the topologies of compact convergence;*

d) *the weak topologies.*

It suffices to apply Proposition 28 exchanging the roles of E and E' and of F and F'.

PROPOSITION 29 *Let E and F be two* LCTVS, u *a continuous linear mapping from E into F. For u to be a homomorphism, it is necessary and sufficient that it be a weak homomorphism (i.e. Proposition 27, that $u'(F')$ be a weakly closed subspace of E') and that every equicontinuous subset of E' contained in $u'(F')$ be the image by u' of an equicontinuous subset of F'.*

Let N be the kernel of u, M its image, v the continuous linear mapping of E/N onto M defined by u. In view of Proposition 20, we verify easily that each of the two hypotheses whose equivalence we wish to prove is equivalent to the same hypothesis on v instead of u. We can thus consider the case where u is a bijection of E onto F and we conclude with the remark that two topologies on an LCTVS are identical if and only if they give the same dual and the same equicontinuous subsets on the dual (Section 11), or also by applying to u Corollary 1 of Proposition 28.

COROLLARY 1 *Let E and F be two* LCTVS, u *a continuous linear mapping from E into F. For u to be an isomorphism, it is necessary and sufficient that every equicontinuous subset of E' be the image by u' of an equicontinuous subset of F'.*

COROLLARY 2 *Let E and F be two* LCTVS, u *a continuous linear mapping from E into F. For u to be a homomorphism from E onto a dense subspace of F it is necessary and sufficient that $u'(F')$ be a weakly closed subspace of E', and that the inverse image by u' of an equicontinuous subset of E' be equicontinuous.*

From this we conclude:

COROLLARY 3 *Let E and F be two* LCTVS, F *being equipped with the Mackey topology $\tau(F, F')$. Let u be a continuous linear mapping from E onto F. For u to be a homomorphism, it is necessary and sufficient that u be a weak homomorphism.*

Using Propositions 29 and 27 we obtain the conditions on u' for u to be a homomorphism from E onto F, etc.

COROLLARY 4   *Let E and F be two LCTVS, u a continuous linear mapping from E onto F, E_0 a dense vector subspace of E. If u induces a homomorphism from E_0 into F, u is a homomorphism.*

In fact, the duals of E_0 and E can be identified with each other; the equicontinuous subsets of one correspond exactly to the equicontinuous subsets of the other; the corollary follows from the fact that a subspace of E' which is closed for $\sigma(E', E_0)$ is a fortiori closed for $\sigma(E', E)$. Note carefully that the converse of Corollary 3 is false: u can be a homomorphism without inducing a homomorphism of E_0 (see Exercise 2).

In the next Section we shall characterize, using the transpose, a *metric* homomorphism from a Banach space into another.

Bitranspose: Let (E, E') and (F, F') be two separated dual systems, u a weakly continuous linear mapping from E into F. Then u' is a continuous mapping from strong F' into strong E' (Proposition 28, Corollary 4), therefore its transpose $(u')'$ is a linear mapping from E'' into F'' continuous for $\sigma(E'', E')$ and $\sigma(E'', F')$ (same reference). It is called the *bitranspose* of u and denoted by u''. It is also the mapping obtained from u by extension by weak continuity to E'' (recall that E is weakly dense in E'') and often it is of interest to reason directly on u'' with this interpretation.

EXERCISE 1   Let E be a vector space, (u_i) a family of linear mappings from E into an LCTVS E_i, for every i let u_i^* be the algebraic transpose of u_i. Let $\mathfrak{S}$ be the set of subsets of the algebraic dual E_i^* of E of the form $u_i^*(A_i)$, where A_i runs through the set of weakly compact equicontinuous disks of E_i'. Show that, on E, the $\mathfrak{S}$-topology is identical to the coarsest topology for which the u_i are continuous. Using Theorem 7, Corollary 1 (Section 12) find a new proof of Theorem 11 (Section 15).

EXERCISE 2   Let E be a metrisable, complete, Hausdorff LCTVS, F a closed vector subspace of infinite codimension. Show that an algebraic supplement E_0 of F exists which is everywhere dense in E, and that for such an E_0, the canonical homomorphism from E onto E/F induces on E_0 a linear mapping which is not a homomorphism.

EXERCISE 3   Let K be a compact space, E an LCTVS, F a closed vector subspace of E. Show that the natural linear mapping from $C(K, E)$ into

$C(K, E/F)$ is a homomorphism from the first space onto a dense subspace of the second. (Use Proposition 29, Corollary 3, replacing $C(K, E)$ by $C(K) \otimes E$, then apply Chapter 1, Section 14, Exercise 2.)

17 Summary and complementary results for normed spaces

Let E be a normed space. Then E' has a natural norm

$$\| x' \| = \sup_{\| x \| \leqslant 1} | \langle x, x' \rangle |,$$

which makes it a Banach space (Chapter 1, Section 5, Theorem 2). We see immediately that the corresponding topology on E' is identical to the strong topology of the dual, i.e. the topology of uniform convergence on the unit ball of E (which is identical to the strong topology, by virtue of the metric characterization of the bounded subsets of E). By definition of the norm of E' the unit ball of E' is the polar of the unit ball of E. Therefore, the unit ball of E' is a *weakly compact* subset of E' ("weak" refers to the topology $\sigma(E', E)$).

The equicontinuous subsets of E' are the bounded subsets of the normed space E', therefore the topology of the bidual E'' of E (Section 12, Definition 9), is identical to the strong topology of the dual of E', i.e. the topology defined by the norm of the dual of the normed space E'. Furthermore:

PROPOSITION 30   *Let E be a normed space. Then the continuous mapping of E into its bidual is a metric isomorphism from E into E''.*

In fact, if $x \in E$ it follows from the definitions that the norm of x in E'' is at most equal to its norm in E; the inverse inequality follows from the stronger fact that there exists an $x' \in E'$ of norm $\leqslant 1$ such that $\langle x, x' \rangle = \| x \|$: it is in fact trivial when we replace E by the line F generated by x, then it is sufficient to extend the linear form obtained on F into a linear form of norm $\leqslant 1$ defined on all of E, by the Hahn--Banach. Theorem II.

COROLLARY   *Every normed space is isomorphic (with its norm) to a vector subspace of a space $C(K)$ constructed on some compact space K.*

We can choose as K the unit ball of E' equipped with the weak topology and then consider for every $x \in E$ the restriction to K of the linear form on E' defined by x; Proposition 30 means precisely that the linear mapping from E into $C(K)$ thus obtained is a metric isomorphism.

Theorem 8 of Section 12 gives

PROPOSITION 31   *Let E be a normed space. E is reflexive if and only if its unit ball is weakly compact.*

In the general case, the bipolar theorem shows that the unit ball of E'' is the weak closure of the unit ball of E ("weak" refers to the topology $\sigma(E'', E')$). We point out the

COROLLARY   *Let E be a reflexive normed space, F a closed vector subspace. Then F and E/F are reflexive.*

It is sufficient to show this for the quotient space. The unit ball of E being weakly compact, its canonical image in E/F is weakly compact, therefore closed, and since it is dense in the unit ball of E/F it is identical to this unit ball. Thus E/F satisfies the criterion of Proposition 31.

In the theory of duality for subspaces and quotients we have

PROPOSITION 32   *Let E be a normed space, F a closed vector subspace. Then the natural linear mapping from the dual of E/F into the dual of E is a metric isomorphism from the first space into the second. The natural linear mapping of the dual of E into the dual of F is a metric homomorphism from the first space onto the second.*

The first assertion follows trivially from the definitions, the second one is already included in the theorem of Hahn–Banach II (therefore it is even an *equivalent* statement if we take into consideration Theorem 6, Section 11). Thus, the canonical isomorphisms $(E/F)' \approx F^0$, $F' \approx E'/F^0$, are even isomorphisms for the normed structure and a fortiori for the various natural strong topologies.

COROLLARY 1   *Let u be a continuous linear mapping from a normed space E into another, F. If u is a homomorphism (resp. metric homomorphism) then its transpose u' is a homomorphism (resp. metric homomorphism).*

In fact, the two parts of Proposition 32 lead us back to the case where u is a *bijection* from E onto F, the rest is trivial.

COROLLARY 2   *Let u be a continuous linear mapping from a complete normed space E into a normed space F. For u to be a metric homomorphism from, E onto F (resp. a metric isomorphism) it is necessary and sufficient that its transpose u' be a metric isomorphism from F' into E' (resp. a metric homomorphism from F' onto E').*

This is just another formulation of Corollary 1. We point out that we

G

shall see in Chapter 4, Part 2, Section 4, that the converses to Corollaries 1 and 2 are valid.

While considering transpositions, it remains to show

PROPOSITION 33 *Let E and F be two normed spaces, u a continuous linear mapping from E into F. Then the norm of u is equal to the norm of its transpose.*

In fact, we have

$$\| u \| = \sup_{\|x\| \leqslant 1} \| ux \|,$$

then by Proposition 30 we get

$$\| ux \| = \sup_{\|x'\| \leqslant |} | \langle ux, x' \rangle |,$$

$$\| u \| = \sup_{\substack{\|x\| \leqslant 1 \\ \|x'\| \leqslant 1}} | \langle ux, x' \rangle | = \sup_{\substack{\|x\| \leqslant | \\ \|x'\| \leqslant |}} | \langle x, u'x' \rangle |$$

$$= \sup_{\|x'\| \leqslant 1} \| u'x' \| = \| u' \|$$

EXERCISE 1 Let E be a normed space, F a closed vector subspace of E. Show that if F and E/F are reflexive, E is reflexive. (Show directly that every continuous linear form on E' comes from an element of E, showing first that it coincides on F^0 with a linear form defined by an element of E.)

EXERCISE 2 Show that a separable normed space is isomorphic to a normed vector subspace of a space $C(K)$, where K is a compact *metrisable* space (use Section 11, Exercise 2). From this conclude that we can choose for K the Cantor set (see Bourbaki, *Topologie Gènèrale*, Chapter IX, Section 2, Exercise 18), or also the interval $I = (0, 1)$ (show that by extending linearly in the components of $\mathbf{C}_I K$ every continuous function given on the Cantor set K, we obtain a metric isomorphism from $C(K)$ into $C(I)$). (Compare these results with Chapter 1, Section 14, Exercise 1.)

EXERCISE 3 Let E be an LCTVS, let u be a continuous linear mapping from E into a Banach space F. Show that the image of the unit ball of F' by the transpose u' of u is an equicontinuous weakly compact disk in E'. Conversely, for every weakly compact equicontinuous disk A' in E', we can find a continuous linear mapping u from E into a Banach space F, such that A' be the image of the unit ball of F' by the transpose u' of u; we can suppose u' bijective. (Consider on E the semi-norm associated to the polar disk of A', and choose F to be the associated Banach space.)

EXERCISE 4

a) Consider the Banach space c_0 of sequences of scalars tending to 0 (see Chapter 1, Section 9, Exercise 7). Let e_i be the element of c_0 with all "coordinates" zero except for the ith which equals 1. Show that the sequence (e_i) tends to zero weakly but not strongly. Show that the sequence

$$u_n = \sum_{1 \leqslant i \leqslant n} e_i$$

is a weak Cauchy sequence in c_0, which does not converge weakly to a limit in c_0 (show that (u_n) converges weakly in the bidual l^∞ of c_0 to the limit u where all the coordinates are equal to 1).

b) Consider the Banach space l^1 of summable sequences of scalars, constructed on a set I of indices. Show that on this space, every weak Cauchy sequence is a strong Cauchy sequence and therefore strongly convergent.

EXERCISE 5   Let E be a separable Banach space where there exist weak Cauchy sequences not weakly convergent or weakly convergent sequences not strongly convergent (see Exercise 4, a)), let u be a homomorphism from l^1 onto E (see Chapter 1, Section 14, Exercise 1). Show that the kernel of u does not admit a topological supplement in l^1 (use Exercise 4, b)).

18 Elementary properties of compactness and weak compactness

In this section we will consider only the most elementary properties of compactness connected with the theory of duality. Further results are contained in Chapter 5.

PROPOSITION 34   *Let A be a subset of an LCTVS E. For A to be precompact it is necessary and sufficient that A be bounded, and that on A the uniform structure induced by E be identical to the induced weak uniform structure.*

The condition is sufficient, since a bounded set is weakly precompact (Section 10, Proposition 15). It is necessary since a precompact set is bounded (Chapter 1, Section 7, Proposition 10); on the other hand, in order to show that the two induced uniform structures are identical, we can clearly suppose E Hausdorff, therefore E can be identified with a space of functions on E' having a $\mathfrak{S}$-convergent topology, the weak topology being the simply convergent topology. Thus we have a general situation already known in topology (see Chapter 0, Proposition 6, Corollary 2).

PROPOSITION 35   *Let A be a precompact subset of an LCTVS E ; then the disked hull of A is still precompact.*

The proof is obvious.

COROLLARY   *Let A be a precompact subset of a Hausdorff LCTVS E. The closed disked hull of A is compact if and only if it is complete (this is the case if in E every bounded closed subset is complete, a fortiori if E is complete).*

In order to verify that a subset B of an LCTVS is complete, the following criterion is often useful:

PROPOSITION 36   *Let E be an LCTVS, A a complete subset of E. Then A is also complete for every locally convex topology T on E which is finer than the initial topology T_0 and which admits a closed (for T_0) fundamental system of neighborhoods of 0.*

The proof reduces to the case where T_0 is Hausdorff, and therefore can be identified with a space of linear forms on E' with a $\mathfrak{S}_0$-topology. Then T is a $\mathfrak{S}$-topology with $\mathfrak{S} \subset \mathfrak{S}_0$, a case known in general topology (Chapter 0, Proposition 6, Corollary 1). We can also prove the more general result: Let F be a set with two uniform structures T_0 and T, T finer than T_0 with a fundamental system of entourages closed in $F \times F$ for the topology corresponding to T_0. Then every subset A of F complete for T_0 is complete for T.

We must show that a Cauchy filter Φ for T admits a limit point for T. By the hypothesis on T, the closures $\bar{B}$, for T_0, in A, $B \in \Phi$, still form a Cauchy filter base Ψ for T; we must show it converges or that

$$\bigcap_{B \in \Phi} \bar{B}$$

is not empty. Since Ψ is also a Cauchy filter base for T_0, formed of closed sets for T_0, and since A is complete for T_0, A admits a cluster point for T, i.e. $\bigcap_{B \in \Phi} \bar{B}$ is not empty.

COROLLARY 1   *Let E be a Hausdorff LCTVS; then every weakly compact subset of E is complete.*

It is in fact weakly complete, and it suffices to apply Proposition 36 with T_0 the weak topology, T the given topology. In view of Proposition 35, we get:

COROLLARY 2   *Let A be a precompact subset of a Hausdorff LCTVS. Its closed disked hull B is compact if and only if B is weakly compact.*

The necessity is clear, the sufficiency follows from Corollary 1 and Proposition 35, Corollary. Finally we point out

COROLLARY 3   *Let E be an* LCTVS. *Every weakly compact subset of E' is strongly complete.*

It is in fact weakly complete, and it suffices to apply Proposition 36 to E' with T_0 the weak topology, T the strong topology.

PROPOSITION 37

1) *Let E be a Hausdorff* LCTVS, *A a compact subset of E, B a closed subset of E. Then $A + B$ is closed.*

2) *Let A, B be two compact convex subsets of E, then $\Gamma(A \cup B)$ is compact.*

1) Let $x \notin A + B$. This also means that the compact set $x - A$ and the closed set B do not meet. It is then well known that there exists a neighborhood $(x - A) + U$ of the compact set which does not meet B; then $x + U$ is a neighborhood of x which does not meet $A + B$.

2) See Section 5, Exercise 4.

THEOREM 12   *Let (E, E') and (F, F') be two dual systems, $\mathfrak{G}$ a set of bounded subsets of E, $\mathfrak{T}$ a set of bounded subsets of F', u a weakly continuous linear mapping from E into F, u' its transpose. The following conditions are equivalent:*

1) *u transforms the $A \in \mathfrak{G}$ into precompact subsets of F equipped with the $\mathfrak{G}$-topology.*

1') *u' transforms the $B \in \mathfrak{T}$ into subsets of E' which are precompact for the $\mathfrak{G}$-topology.*

2) *The restrictions of u to the $A \in \mathfrak{G}$ are uniformly continuous when we equip E with the weak topology and F with the $\mathfrak{G}$-topology.*

2') *The restrictions of u' to the $B \in \mathfrak{T}$ are uniformly continuous when we equip F' with the weak topology and E' with the $\mathfrak{G}$-topology.*

3) *The restriction of the function $\langle ux, y' \rangle = \langle x, u'y' \rangle$ to the sets $A \times B$ with $A \in \mathfrak{G}$, $B \in \mathfrak{T}$ are uniformly continuous for the product of the weak uniform structures. It even suffices to suppose these functions to be uniformly continuous for the product of the weak uniform structure of A by the strong uniform structure of B, or conversely.*

Furthermore, when the $A \in \mathfrak{G}$ (resp. the $B \in \mathfrak{T}$) are disked we can replace in Condition 2 (resp. 2') the uniform continuity by continuity and even by *continuity at the origin. The preceding conditions imply that for every $A \in \mathfrak{S}$, $B \in \mathfrak{T}$, the set of $u(x, y')$, with $x \in A$ $y' \in B$ is bounded.*

Proof 1) implies 2'). Since the $B \in \mathfrak{T}$ are uniformly continuous sets of functions on F (with $\mathfrak{S}$-topology), on such a B the weak uniform structure is identical to the structure of precompact convergence and, also, 1) means precisely that u' is continuous, therefore uniformly continuous, for precompact convergence on F' and $\mathfrak{S}$-convergence on E' (Section 16, Proposition 28). 2') implies 1'), by reasons of uniform continuity since the $B \in \mathfrak{T}$ are weakly precompact subsets of F'. By reasons of symmetry we also have the implications: 1') $\Rightarrow$ 2) and 2') $\Rightarrow$ 1), which proves that conditions 1), 2'), 1'), 2) are equivalent.

Furthermore, 2) and 2') imply 3) as can easily be seen by writing

$$\langle ux_1, y_1' \rangle - \langle ux_2, y_2' \rangle = \langle u(x_1 - x_2), y_1' \rangle + \langle ux_2, y_1' - y_2' \rangle,$$

and conversely, the first weakened statement of 3) evidently implies 2), since it implies that for a given $\varepsilon > 0$ there exists a weak entourage U in A such that

$$| \langle ux_1, y' \rangle - \langle ux_2, y' \rangle | \leqslant \varepsilon$$

for $x_1, x_2 \in A$, $y' \in B$, $(x_1, x_2) \in U$ i.e. $ux_1 - ux_2 \in B^0$ which is precisely 2). Thus 3) and its variants are equivalent to the preceding conditions. Since $A \times B$ is precompact for the product of the weak uniform structures, 3) implies that the set of $\langle ux, y' \rangle$, with $x \in A$, $y' \in B$ is a precompact subset, therefore bounded, of the field of scalars. Finally, the substitution in 2) (or 2')) of uniform continuity by continuity at the origin is a particular case of Section 14, Lemma.

COROLLARY 1 *Let u be a continuous linear mapping from an LCTVS E into another, F, and $\mathfrak{S}$ a set of bounded subsets of E that generate E. The mapping u transforms the $A \in \mathfrak{S}$ into precompact subsets of F if and only if its transpose u' transforms the equicontinuous subsets of F' into relatively compact subsets of E' equipped with the $\mathfrak{S}$-topology.*

In fact, in the second condition we can suppose that the equicontinuous subset B of F' is weakly compact, then its image in E' is weakly compact therefore weakly complete and a fortiori complete for $\mathfrak{S}$-convergence (Proposition 37); it follows that precompactness of B is equivalent to relative compactness. Then, Corollary 1 is a particular case of the equivalence of conditions 1) and 1') of Theorem 12. Notice that we can also see that, if the $A \in \mathfrak{S}$ are weakly compact and F Hausdorff, then the condition considered in Corollary 1 means that u transforms the $A \in \mathfrak{S}$ into *relatively compact* subsets of F. A simple change in notation gives us:

COROLLARY 2   *Let E and F be two Hausdorff LCTVS, u a weakly continuous linear mapping from E' into F, u' the weakly continuous linear mapping from F' into E (u' is the transpose of u). u transforms the equicontinuous subsets of E' into relatively compact subsets of F if and only if u' does.*

We remark that the first condition means also that u' is continuous when F' has the topology of uniform convergence on the compact disks of F (Section 16, Proposition 28); this is equivalent to an analogous statement on u.

COROLLARY 3   *Let u be a continuous linear mapping from a Banach space E into another, F. The mapping u transforms the unit ball of E into a relatively compact subset of F if and only if its transpose u' transforms the unit ball of F' into a relatively compact subset of the Banach space E'.*

(Choose $\mathfrak{S}$ with only the unit ball of E, $\mathfrak{T}$ with only the unit ball of F'.)

COROLLARY 4   *Let (E, E') be a dual system, $\mathfrak{S}$ a set of bounded subsets of E, $\mathfrak{T}$ a set of bounded subsets of E'. The following conditions are equivalent:*

a) *The subsets $A \in \mathfrak{S}$ are precompact for the $\mathfrak{S}$-topology;*

a') *The subsets $A' \in \mathfrak{T}$ are precompact for the $\mathfrak{S}$-topology;*

b) *For every $A \in \mathfrak{S}$, $A' \in \mathfrak{T}$, the restriction of the function $\langle x, x' \rangle$ to $A \times A'$ is uniformly continuous for the product of the weak uniform structures.*

It suffices to apply Theorem 12 to the identity mapping of E. In particular, when $\mathfrak{S}$ is the set of bounded subsets of an LCTVS E, $\mathfrak{T}$ the set of equicontinuous subsets of E', we obtain

COROLLARY 5   *Let E be an LCTVS. The bounded subsets of E are precompact if and only if the equicontinuous subsets of E' are strongly relatively compact.*

This is also a particular case of Corollary 1, relative to the identity mapping of E and the set of all bounded subsets of E.

DEFINITION 13   *Let E be an LCTVS. E is a Montel space, abbreviated a space of type $(\mathscr{M})$, if it is Hausdorff and if every bounded subset of E is relatively compact.*

A fortiori, such a space is reflexive. We have seen that the spaces $\mathscr{E}(U)$ and $\mathscr{E}(K)$ of Schwartz (Chapter 1, Section 10) are Montel spaces.

A Banach space of type ($\mathcal{M}$) is finite dimensional by Chapter 1, Section 13. Corollary 4 of Theorem 12 then gives

PROPOSITION 38   *Let E be a Hausdorff LCTVS. E is of type ($\mathcal{M}$) if and only if its closed bounded subsets are complete and the equicontinuous subsets of E' strongly relatively compact.*

COROLLARY   *Let E be an LCTVS whose closed bounded subsets are complete and such that the strongly bounded subsets of the dual are equicontinuous (example, a complete metrisable space—see Chapter 1, Section 15, Theorem 11). E is of type ($\mathcal{M}$) if and only if its strong dual is of type ($\mathcal{M}$).*

(We shall study in Chapter 3, Section 3, under the name of quasi-barrelled spaces, the LCTVS such that the strongly bounded subsets of the dual are equicontinuous.)

DEFINITION 14   *Let u be a linear mapping of an LCTVS E into another, F. We say that u is compact (resp. precompact, resp. weakly compact, resp. bounded) if u transforms some neighborhood V of 0 in E into a relatively compact resp. precompact, . . .) subset of F.*

When E is a Banach space, we may choose for V the unit ball of E. To say that u is bounded is to say that u is continuous (Chapter 1, Section 7, Proposition 12, Corollary 2). Corollary 3 of Theorem 12 states that a linear mapping u from a Banach space into another is compact if and only if its transpose is compact.

We now examine weak compactness. Recall that as the bounded subsets of an LCTVS E are exactly the weakly precompact subsets, a subset A of E is weakly compact if and only if it is bounded, weakly Hausdorff and weakly complete.

The analogue of Theorem 12 is

THEOREM 13   *Let u be a continuous linear mapping from a Hausdorff LCTVS E into another, F. Let $\mathfrak{S}$ be a set of bounded subsets in E, let H be the vector space generated by the weak closures of the $A \in \mathfrak{S}$ in E''; we suppose $H \supset E$. The following conditions are equivalent:*

1) u transforms the $A \in \mathfrak{S}$ into weakly relatively compact subsets of F

2) The bitranspose u'' of u maps H into F.
These conditions imply

3) The transpose u' transforms the weakly closed equicontinuous subsets of F' into subsets of E' which are relatively compact for $\sigma(E', H)$.

When F is quasi-complete, Condition 3) also implies Conditions 1) and 2).

Proof   Since u'' is a weakly continuous mapping from E'' into F'' and since the weak closures in E'' of the $A \in \mathfrak{S}$ are weakly compact sets, 1) is equivalent to 2). Also, 2) means that u' is continuous for $\sigma(F', F)$ and $\sigma(E', H)$ and, since an equicontinuous weakly closed subset B of F' is weakly compact, 2) implies that B is transformed into a compact, therefore relatively compact subset of E' with $\sigma(E', H)$. Conversely, if this is so, the topology $\sigma(E', H)$ on $u'(B)$ will be identical to $\sigma(E', E)$, therefore the restriction of u' to an equicontinuous subset B of F' will be continuous for the topologies $\sigma(F', F)$ and $\sigma(E', H)$, therefore for every $x \in H$ the restriction of the linear form $u''x = x \circ u'$ to the equi-continuous subsets of F being weakly continuous, belongs to the completion of F (Section 14, Theorem 10, Corollary 2), and therefore to F when F is complete. In any case, $u''x$ belongs also to the weak closure in the completion $\hat{F}$ of F, of a bounded subset of F. We can obviously suppose this subset disked, therefore its weak closure in $\hat{F}$ is identical to its closure for the natural topology of the completion; then, if F is quasi-complete, we will have $u''x \in F$ for every $x \in H$.

The most important case is the one where $\mathfrak{S}$ is the set of all bounded subsets of E:

COROLLARY 1   *Let E, F be two Hausdorff LCTVS, u a continuous linear mapping from E into F. The following conditions are equivalent:*

1) *u transforms the bounded subsets of E into weakly relatively compact subsets of F;*

2) *the bitranspose u'' maps E'' into F.*

These conditions imply:

3) *The transpose u' transforms the equicontinuous subsets of F' into relatively compact subsets of E' for $\sigma(E', E'')$; the converse is true if F is quasi-complete.*

COROLLARY 2   *Let E, F be Hausdorff LCTVS, u a weakly continuous linear mapping from E into F', u' the weakly continuous mapping from F into E', u' the transpose of u. The mapping u transforms the bounded subsets of E into relatively compact subsets of F' for $\sigma(F', F'')$ if and only if u' has the analogous property.*

Proof   It suffices to show that the first condition implies the second, which follows from Corollary 1, applied to E and F' strong, since the bounded subsets of F are equicontinuous subsets of the dual of F' strong.

COROLLARY 3   *Let u be a continuous linear mapping from a Banach space E into another, F. The mapping u is weakly compact (see Definition*

14) *if and only if its transpose u' is a weakly compact mapping from the Banach space F' into the Banach space E'.*

COROLLARY 4   *Let E be a Hausdorff* LCTVS.

a) *If E is reflexive and if the strongly bounded subsets of the dual are equicontinuous, then E' equipped with the strong topology is reflexive.*

b) *If E is quasi-complete and E' with the strong topology is reflexive, then E is reflexive.*

c) *Thus, if the strongly bounded subsets of E' are equicontinuous, E is reflexive if and only if it is quasi-complete and its strong dual is reflexive. (In particular, a metrisable and complete* LCTVS *is reflexive if and only if its strong dual is reflexive.)*

a) is an immediate consequence of Theorem 8 and b) results from Corollary 1 for the identity mapping.

EXERCISE 1   Let A, B be two sets, u a bounded scalar function on $A \times B$. For every $x \in A$ let $\tilde{x}$ be the function $y \mapsto u(x, y)$ on B and for every $y \in B$ let $\tilde{y}$ be the function $x \mapsto u(x, y)$ on A. Let $\tilde{A}$ be the subset of the Banach space $C^{\infty}(B)$ of bounded functions on B formed by the $\tilde{x}(x \in A)$ and let $\tilde{B}$ be the analogous subset of $C^{\infty}(A)$. Show that $\tilde{A}$ is a relatively compact (or weakly relatively compact) subset of $C^{\infty}(B)$, if and only if $\tilde{B}$ is a relatively compact (or a weakly relatively compact) subset of $C^{\infty}(A)$.

Application   Let G be a monoid, f a bounded function on G, for every $s \in G$ let $U_s f$ (or $V_s f$) be the function $t \mapsto f(st)$ (or $t \mapsto f(ts)$) on G. Show that the set of left translates $U_s f$ of f is relatively compact (or weakly relatively compact) in $C^{\infty}(G)$ if and only if the set of right translates $V_s f$ is. We say that f is an almost periodic function (or weakly almost-periodic) on G. Show that if f is almost periodic then the set of $U_s V_t f$ where $s, t \in G$ is a relatively compact subset of $C^{\infty}(G)$.

EXERCISE 2   Consider the Banach spaces c_0, l^1, l^{∞} (Chapter 1, Section 9, Exercises 6, 7).

a) Show that c_0, l^1 are separable and from this conclude that the unit ball of c_0 is weakly metrisable, that the unit ball of l^1 is metrisable for $\sigma(l^1, c_0)$. Show that l^{∞} is not separable (for every subset A of the set of the integers, let $x_A \in l^{\infty}$ be its characteristic function, show that $A \neq B$ implies $\| x_A - x_B \| \geqslant 1$ and that the set of x_A is not countable.

b) Show that every weakly compact subset of the Banach space l^1 is compact (use Section 17, Exercise 4b; it suffices to show that from every weakly relatively compact sequence of l^1 we can extract a sub-sequence that converges strongly, using a) in order to extract a sequence that converges for $\sigma(l^1, c_0)$ and therefore for $\sigma(l^1, l^\infty)$).

c) Let u be a continuous linear mapping from c_0 into an LCTVS E; show that u is compact if and only if u is weakly compact (use b)).

d) Generalize b) and c) to the spaces $l^1(I)$ etc. constructed on an index set I.

EXERCISE 3 Let E be a Hausdorff LCTVS, u a weakly continuous linear mapping from the dual l^∞ of l^1 into E (u therefore transforms the unit ball of l^∞ into a weakly compact subset of E). Show that u transforms even the unit ball of l^∞ into a compact subset of E (see Exercise 2). Let e_i be the element of l^∞ with all coordinates zero except the ith which equals 1; show that the sequence of ue_i is a commutatively convergent series in E, that its product with every bounded sequence is also and that

$$u((x_i)) = \sum x_i ue_i$$

for every $(x_i) \in l^\infty$. Conversely, show that if E is quasi-complete every summable family in E is obtained in this way. Therefore there is a bijective correspondence between the weakly continuous linear mappings from l^∞ into E, the compact linear mappings (or weakly compact) from c_0 into E and the commutatively convergent series in E. (The spaces c_0, l^1, l^∞ can still be constructed on any given set I of indices.)

EXERCISE 4

a) Let u be a bilinear form on the product $E \times F$ of two LCTVS, let A (or B) be a disk in E (or F). Show that if the restriction of u to $A \times B$ is continuous at the origin, then the restriction is continuous. If furthermore A is precompact and B compact, then u is continuous on $A \times B$.

b) Let u be a weakly continuous linear mapping from an LCTVS E into another, F; let $\mathfrak{S}$ be a set of bounded disks of E covering E. The mapping u transforms the subsets $A \in \mathfrak{S}$ into precompact subsets of F if and only if for every $A \in \mathfrak{S}$ and every equicontinuous subset B of F', the restriction to $A \times B$ of the function $\langle ux, y' \rangle$ is continuous at the origin for the product of the weak topologies (see Theorem 12, Conditions 1) and 3)).

c) Let E be a separable Banach space. Show that for every sequence (x_i) in E converging weakly to 0 and every sequence (x_i') in E' converging weakly to 0 we have

$$\langle x_i, x_i' \rangle \to 0,$$

if and only if the weakly compact subsets of E are compact (use b), by noticing that the unit ball of E' is weakly metrisable, that consequently E' is weakly separable and the weakly compact subsets of E are weakly metrisable.

EXERCISE 5   Let E be an LCTVS whose strong dual is separable (for example c_0, see Exercise 2a, or a vector subspace of c_0).

a) Let u be a weakly continuous linear mapping from E into an LCTVS F. The mapping u transforms the bounded subsets into precompact subsets if and only if u transforms the sequences that converge weakly to 0, into convergent sequences for the given topology (it suffices to see that the restriction of u to a bounded set A of E is continuous for the weak topology of E and the given topology of F and to note that A is weakly metrisable).

b) Let F be an LCTVS whose weakly compact subsets are compact. Show that every weakly continuous linear mapping from E into F transforms the bounded subsets into precompact subsets (and is therefore precompact if E is a normed space). (In particular we can choose E to be a vector subspace of c_0 and $F = l^1$; see Exercise 2).

c) From this, conclude that if E is a normed space whose strong dual is separable and whose weakly compact subsets are compact, then E is finite dimensional.

d) Conclude from c) and from Exercise 4c, that if E is an infinite dimensional Banach space whose strong dual is separable we can find in E a sequence (x_i) converging weakly to 0, such that $\langle x_i, x_i' \rangle = 1$ for every i.

e) If E is an LCTVS whose dual is separable, show that every element of E'' is the limit of a weak Cauchy sequence in E (notice that the bounded subsets of E are weakly metrisable). From this conclude that a weakly continuous linear mapping u from E into an LCTVS F transforms the bounded subsets into weakly relatively compact subsets if and only if u transforms weak Cauchy sequences into weakly convergent sequences.

CHAPTER 3

Spaces of linear mappings

IN THIS CHAPTER we continue to develop the formalism begun in the preceding one, concentrating on spaces of linear mappings. The development is easy and we do not meet any truly new theorems; the hypothesis of local convexity (and the Hahn–Banach theorem) are seldom used. Mainly we use General Topology and the application of the Banach–Steinhaus theorem (therefore of Baire's theorem).

1 Generalities on the spaces of linear mappings

Let E and F be LCTVS, let $L(E, F)$ be the space of continuous linear mappings from E into F. Let y be an element of F not in the closure of the origin, i.e. such that the line generated by y is Hausdorff therefore isomorphic to the field of scalars. If we assign to every $x' \in E'$ the mapping

$$x \mapsto \langle x, x' \rangle y$$

from E into F we clearly obtain an algebraic isomorphism from E' into $L(E, F)$. If $\mathfrak{S}$ is any given set of subsets of E, the preceding isomorphism is also a homomorphism from E' into $L(E, F)$ when these spaces are equipped with the $\mathfrak{S}$-topology. It follows that (in view of Chapter 2, Section 10, Proposition 16) the $\mathfrak{S}$-topology on $L(E, F)$ is compatible with the vector structure only if the $A \in \mathfrak{S}$ are bounded subsets of E. The condition is also sufficient by the general criterion of Chapter 1, Section 8, Theorem 3. In what follows, when we consider on $L(E, F)$ a $\mathfrak{S}$-topology, $\mathfrak{S}$ will always be a set of *bounded* subsets of E.

Let $L_{\mathfrak{S}}(E, F)$ be the space $L(E, F)$ of continuous linear mappings from E into F equipped with the $\mathfrak{S}$-topology. In particular, $L_s(E, F)$ and $L_b(E, F)$ stand for the space $L(E, F)$ with, respectively, the topology of pointwise convergence and the topology of uniform convergence on the bounded subsets of E. If for every $A \in \mathfrak{S}$ and every neighborhood V of 0 in F we let $W(A, V)$ be the set of $u \in L(E, F)$ such that $u(A) \subset V$, the sets $W(A, V)$ and the homothetics of the intersections of a finite number of such sets, form a fundamental system of neighborhoods of 0 in $L_{\mathfrak{S}}(E, F)$. It follows that the $\mathfrak{S}$-topology does not change when we add to $\mathfrak{S}$ the closed disked hulls of finite unions of elements of $\mathfrak{S}$, the homothetics of such hulls and finally all sets contained in the sets of the preceding type. However, when F is not identical to the closure of the origin, the immersion of E' into $L(E, F)$ indicated at the beginning shows that we cannot increase the set of subsets further without making the corresponding topology finer (Chapter 2, Section 10, Proposition 17). Finally we remark that in the study of a $\mathfrak{S}$-topology we can always suppose $\mathfrak{S}$ to be a set of closed disks of E, closed under homothetics, containing together with a finite number of disks A_i their closed disked hulls and with a disk A all the disks contained in A.

We can also consider the space of all *weakly* continuous linear mappings from E into F with a $\mathfrak{S}$-topology (for the given topology on F); this space is identical to $L_\mathfrak{S}(E_\tau, F)$ where E_τ is E with the Mackey topology $\tau(E, E')$ (since the weakly continuous linear mappings from E into F are precisely those which are continuous for the topology $\tau(E, E')$ on E and the given topology on F; see Chapter 1, Section 16, Proposition 28, Corollary 2). More generally, consider two dual systems (E, E') and (F, F'), let $\mathfrak{S}$ be a set of bounded subsets of E, $\mathfrak{T}$ a set of bounded subsets of F' ("bounded" meaning weakly bounded). We can equip F with a $\mathfrak{T}$-topology, which induces on the space of *weakly* continuous linear mappings from E into F a $\mathfrak{S}$-topology; it is locally convex if for example the $A \in \mathfrak{S}$ are *strongly* bounded or the $B \in \mathfrak{T}$ strongly bounded as we can easily verify. In either case:

PROPOSITION 1   *Let (E, E') and (F, F') be two separated dual systems, let $\mathfrak{S}$ (or $\mathfrak{T}$) be a set of bounded subsets of E (or F). Equip E' with a $\mathfrak{S}$-topology, F with a $\mathfrak{T}$-topology, then equip the space of weakly continuous linear mappings from E into F with the $\mathfrak{S}$-topology, and the space of weakly continuous linear mappings from F' into E' with the $\mathfrak{T}$-topology, then the operation of transposition is homomorphism from the first space of linear mappings onto the second.*

Proof   From the characterization of neighborhoods of 0 in the spaces under consideration it suffices to show that for given $A \in \mathfrak{S}$, $B \in \mathfrak{T}$ the relations $u(A) \subset B^0$ and $u'(B) \subset A^0$ for a weakly continuous linear mapping u from E into F (with transpose u') are equivalent. Since the first relationship can be written $A \subset u^{-1}(B^0)$ and since $u^{-1}(B^0) = (u'(B))^0$ (Chapter 2, Section 16, Proposition 25); from the bipolar theorem, $A \subset (u'(B))^0$ is equivalent to $A^0 \supset u'(B)$.

COROLLARY   *Let (E, E') and (F, F') be two separated dual systems. Con-*

sider on E, E', F, F' the weak topologies. Then the operation of transposition $u \mapsto u'$ is a TVS *isomorphism from $L_s(E, F)$ onto $L_s(F', E')$.*

PROPOSITION 2 *Set E and F be two non-zero* LCTVS. *We equip E with the topology $\tau(E, E')$ so that the linear mappings from E into F which are continuous or weakly continuous are the same, and we suppose F to be Hausdorff. Let $\mathfrak{S}$ be a set of bounded subsets of E such that the vector space generated by their union is identical to E. Then the space $L_\mathfrak{S}(E, F)$ is complete if and only if F is complete and E' is complete for the $\mathfrak{S}$-topology.*

The necessity follows as E' with $\mathfrak{S}$-convergence is isomorphic to a closed topological vector subspace of $L(E, F)$ (see the beginning of this section), and also as F is isomorphic to a closed topological vector subspace of $L_\mathfrak{S}(E, F)$ (choose $x' \in E'$ non-zero and associate with every $y \in F$ the mapping $x \mapsto \langle x, x' \rangle y$ from E into F). Conversely, if E' (with $\mathfrak{S}$-convergence) and F are complete we shall show that $L_\mathfrak{S}(E, F)$ is complete. Since the space of *all* mappings from E into F is complete for $\mathfrak{S}$-convergence, it suffices to show that $L_\mathfrak{S}(E, F)$ is a *closed* subspace, i.e. that every mapping u from E into F which is a limit for the $\mathfrak{S}$-convergence of continuous linear mappings is linear and continuous. The linearity is trivial in any case. For the continuity it is sufficient to verify weak continuity, that is, for every $y' \in F'$, $y' \circ u$ is a continuous linear form on E. It is immediate that $y' \circ u$ is a limit for the $\mathfrak{S}$-convergence of linear forms of type $y' \circ u_i$ where the u_i are in $L(E, F)$, therefore it is a limit for the $\mathfrak{S}$-convergence of continuous linear forms. Since E' is complete for $\mathfrak{S}$-convergence, it follows that $y' \circ u$ is continuous.

Recall that we have obtained in Chapter 2, Section 14 a criterion for the completeness of E' and F (with $\mathfrak{S}$-convergence) which can be used in Proposition 2.

EXERCISE Let E and F be two Hausdorff LCTVS, let $\mathfrak{S}$ be a set of weakly compact disks in E whose union generates E and let F_s be the space F with the weak topology. Show that the dual of $L_\mathfrak{S}(E, F_s)$ is identical to $E \otimes F'$ (use Proposition 1 and Chapter 2, Section 15, Proposition 23). Show that we obtain the same dual by equipping $L(E, F_s) = L(E_\tau, F)$ with the l.u.b. topology of the preceding topology and the topology of the space $L_s(E_\tau, F)$ (use Chapter 2, Section 15, Theorem 11).

2 Bounded sets in the spaces of linear mappings

DEFINITION 1 *Let E be an LCTVS, $\mathfrak{S}$ a set of bounded subsets of E. A subset U of E is $\mathfrak{S}$-absorbing if for every $A \in \mathfrak{S}$ we have $\lambda U \supset A$ for λ positive and sufficiently large. When $\mathfrak{S}$ is the set of all bounded subsets of E, we simply say that U is borniverous.*

When U or the $A \in \mathfrak{S}$ are balanced, it suffices in the preceding definition to suppose that there exists for every $A \in \mathfrak{S}$, a $\lambda \geqslant 0$ such that $\lambda U \supset A$. Notice that if $\mathfrak{S}$ is the set of subsets of E reduced to a point, to say that U is $\mathfrak{S}$-absorbing is to say that U is *absorbing*.

PROPOSITION 3 *Let E and F be LCTVS, $\mathfrak{S}$ a set of bounded subsets of E, M a set of (linear mappings from E into F not necessarily continuous). $M(A)$ is a bounded subset of F for every $A \in \mathfrak{S}$ if and only if for every neighborhood V of 0 in F, the set $M^{-1}(V)$ is a $\mathfrak{S}$-absorbing subset of E.*

(Recall that $M(A) = \bigcup_{u \in M} u(A)$, $M^{-1}(V) = \bigcap_{u \in M} u^{-1}(V)$). The proof is trivial.

COROLLARY 1 *Let E and F be two LCTVS, $\mathfrak{S}$ a set of bounded subsets of E. For a subset M of $L_{\mathfrak{S}}(E, F)$ the following conditions are equivalent:*

a) *M is bounded;*

b) *for every $A \in \mathfrak{S}$, $M(A)$ is a bounded subset of F;*

c) *For every neighborhood of 0 in F, $M^{-1}(V)$ is a $\mathfrak{S}$-absorbing subset of E.*

The equivalence of a) and b) is a particular case of Chapter 1, Section 8, Proposition 15; the equivalence of b) and c) is a particular case of Proposition 3.

COROLLARY 2 *Let E be an LCTVS with a set $\mathfrak{S}$ of bounded sets. A subset A' of E' is bounded for the $\mathfrak{S}$-topology if and only if its polar A'^0 is a $\mathfrak{S}$-absorbing subset of E.*

This is also equivalent to the fact that the weakly closed disked hull of A' is bounded for the $\mathfrak{S}$-topology; furthermore, Corollary 2 shows that there is a bijection, by polarity, between the closed $\mathfrak{S}$-absorbing disks in E, and the weakly closed disks of E' which are bounded for $\mathfrak{S}$-convergence. From this we conclude

COROLLARY 3 *Let E be an LCTVS, let $\mathfrak{S}_1$ and $\mathfrak{S}_2$ be sets of bounded subsets of E. Then the following conditions are equivalent:*

a) *Every $\mathfrak{S}_1$-absorbing closed disk in E is $\mathfrak{S}_2$-absorbing.*

b) *Every subset bounded for $\mathfrak{S}_1$-convergence is bounded for $\mathfrak{S}_2$-convergence, in E'.*

c) *Every $A \in \mathfrak{S}_2$ is bounded for the $\mathfrak{S}'_1$-topology, where $\mathfrak{S}'_1$ is the set of subsets of E' bounded for $\mathfrak{S}_1$-convergence;*

d) *For every LCTVS F, every subset of $L(E, F)$ bounded for $\mathfrak{S}_1$-convergence is bounded for $\mathfrak{S}_2$-convergence.*

a) and b) are equivalent by polarity; a) implies d) by Corollary 1, and b) is a particular case of d), therefore a), b), c) are equivalent. Finally, b) means also (by Proposition 3) that for every subset $A' \in \mathfrak{S}'_1$ of E' and every subset $A \in \mathfrak{S}_2$ of E, the set $\langle A, A' \rangle$ of scalar products $\langle x, x' \rangle$ with $x \in A$, $x' \in A'$ is bounded; this is also equivalent to c). Notice that condition b) is already verified if we suppose d) valid for an LCTVS F which is different from the closure of the origin as in the immersion of E' into $L(E, F)$ described at the beginning of Section 1. From Corollary 3, if $\mathfrak{S}_1$ is a set of subsets of E, there exists a larger set $\mathfrak{S}_2$ of bounded subsets of E such that for every LCTVS F, every subset of $L(E, F)$ bounded for $\mathfrak{S}_1$-convergence is bounded for $\mathfrak{S}_2$-convergence: it is the set of subsets of E' bounded for $\mathfrak{S}_1$-convergence. In particular, when $\mathfrak{S}_1$ is the set of subsets reduced to a point of E (the most interesting case), the corresponding topology on E' being the weak topology, we obtain for $\mathfrak{S}_2$ the set of *strongly bounded* subsets of E (recall that we call the topology of uniform convergence on the weakly bounded subsets of E' the strong topology on E, a topology that depends only on the dual system (E, E')). In particular,

PROPOSITION 3' *Let E and F be LCTVS; every subset of $L(E, F)$ bounded for pointwise convergence, is bounded for uniform convergence on the strongly bounded subsets of E.*

The following theorem reveals an important class of strongly bounded subsets of E:

THEOREM 1 (BANACH–STEINHAUS–MACKEY) *Let E be a Hausdorff LCTVS. Every bounded complete disk of E is strongly bounded.*

In other words:

COROLLARY 1 *Let E and F be LCTVS, E Hausdorff; then every subset of $L(E, F)$ bounded for pointwise convergence is bounded for the uniform convergence on the bounded complete disks of E.*

H

This means that in E every closed and absorbing disk is $\mathfrak{G}$-absorbing, where $\mathfrak{G}$ is the set of complete bounded disks of E. We point out that Theorem 4' of Chapter 2, Section 8, is a particular case of Theorem 1: the case where a Hausdorff LCTVS E is considered as the dual of E' weak for the topology of uniform convergence on some set $\mathfrak{G}$ of weakly compact (therefore weakly complete) disks of E': to say that a subset of E is weakly bounded is to say that it is bounded for the topology of pointwise convergence, therefore, by Theorem 1, it is bounded for $\mathfrak{G}$-convergence. More generally

COROLLARY 2   *Let E and F be LCTVS, E Hausdorff. Let M be a set of continuous linear mappings from E into F bounded for pointwise convergence; then M is bounded for the topology of uniform convergence on the weakly compact disks of E.*

Proof   We can replace the topology of E by the weak topology; a weakly compact subset of E being a fortiori bounded and weakly complete, and apply Theorem 1.

COROLLARY 3   *Let E be a Hausdorff LCTVS, complete, or more generally, whose closed bounded subsets are complete. Then for every LCTVS F, every set of continuous linear mappings from E into F, bounded for the topology of pointwise convergence, is bounded for the topology of bounded convergence.*

DEFINITION 2   *E is quasi-complete if its bounded closed subsets are complete.*

Thus, if E is quasi-complete, the subsets of $L(E, F)$ bounded for the various $\mathfrak{G}$-topologies ($\mathfrak{G}$ is a set of bounded subsets of E whose union generates E) are identical.

Proof of Theorem 1.   We must show that if M is a set of continuous linear mappings from E into F, bounded for pointwise convergence, and if A is a complete bounded disk in E, then $M(A)$ is a bounded subset of F. Now, let E_A be the vector space generated by A, with the *gauge* semi-norm of A:

$$\| x \|_A = \inf_{x \in \lambda A} | \lambda |;$$

this is a norm since A is bounded. Taking the restrictions of $u \in M$ to E_A, we can show that the set of mappings from E_A into F thus obtained, is bounded for A-convergence. But as this set is bounded for pointwise convergence, it suffices to show that E_A is complete and to apply the Banach–Steinhaus theorem (Chapter 1, Section 15, Theorem 11). We then have the following lemma, interesting in itself:

LEMMA 1 *Let E be a Hausdorff LCTVS, A a complete bounded disk in E. Then the corresponding normed space E_A is a Banach space, i.e. it is complete.*

Proof As A is closed in E, and therefore contains the ends of the intervals intersected by A on the real lines passing through the origin, we conclude that the unit ball of E_A is A. Also it is clear that a normed space is complete if and only if its unit ball is complete. It is then sufficient to show that the unit ball A of E is complete for the norm topology of E_A. This follows from Chapter 2, Section 18, Proposition 35, applied to E_A and to the topology induced by E on E_A.

EXERCISE 1 Let E be a Hausdorff LCTVS, A a complete convex (not necessarily disked) subset of E. Show that every set of continuous linear mappings from E into an LCTVS F bounded for pointwise convergence, is bounded for A-convergence. (Examine the case where $0 \in A$ and proceed by contradiction: if M were not bounded on A, there would exist a sequence (x_n) extracted from A such that M would not be bounded on the sequence of x_n/n; consider then the closed convex hull K of the sequence of x_n/n and notice that $K-K$ is a symmetric compact, convex subset of E on which M would not be bounded.)

EXERCISE 2 A sequence of scalars (λ_i) is *rapidly decreasing* if its product with every monomial sequence $i \mapsto i^n$ is bounded. Let E be an LCTVS, a sequence (x_i) in E is rapidly decreasing if for every continuous seminorm p on E, the sequence of $p(x_i)$ is rapidly decreasing. Show that this is true if and only if for every integer $n > 0$, the sequence $(i^n x_i)$ is bounded. From this conclude that the sequence (x_i) in E is rapidly decreasing if and only if for every $x' \in E'$, the sequence $(\langle x_i, x' \rangle)$ is rapidly decreasing. Generalize this result for a larger class of sequences.

EXERCISE 3

a) Let u be a continuous linear mapping from an LCTVS E into another, F. Show that u is continuous for the strong topologies (apply Chapter 2, Section 16, Proposition 28, Corollary 4) and therefore transforms the strongly bounded subsets into strongly bounded subsets. In particular, if u is a continuous linear mapping from a Banach space E into an LCTVS F, then u transforms the unit ball of E into a strongly bounded subset of F, i.e. it is continuous for the strong topology of F.

b) Let F be an LCTVS with the topology T, let T' be another LC topology on F with a fundamental system of neighborhoods of 0 closed for T. Show that every linear mapping u from a Banach space E into F continuous for T, is continuous for T' (notice that the topology T

is the topology of uniform convergence on a set of weakly bounded subsets of the dual F' of F with the topology T).

c) Particular case: let E and F be LCTVS, u a continuous linear mapping from a Banach space H into $L_s(E, F)$ then u is also continuous from H into $L_b(E, F)$.

EXERCISE 4   Let (E, E') be a dual system. Every bounded subset of E is strongly bounded if and only if the same is true for E'.

3 Relationship between bounded sets and equicontinuous sets. Barrelled spaces

Let E and F be LCTVS, M a set of linear mappings from E into F. The set M is equicontinuous if and only if for every neighborhood V of 0 in F, the set $M^{-1}(V)$ is a neighborhood of 0 in E. Notice that if M is equicontinuous, then the sets $M^{-1}(V)$ are a fortiori $\mathfrak{S}$-absorbing for every set $\mathfrak{S}$ of bounded subsets of E, therefore by Proposition 3:

PROPOSITION 4   *Let E and F be* LCTVS, *M an equicontinuous set of linear mappings from E into F. Then M is bounded for every $\mathfrak{S}$-topology ($\mathfrak{S}$ is a set of bounded subsets of E).*

We have already seen this proposition in Chapter 1, Section 15, Proposition 22. We examine the case where there is a converse. First, consider the

LEMMA   *Let E be an* LCTVS, *$\mathfrak{S}$ a set of bounded subsets of E. The following conditions are equivalent:*

a) *In E every closed $\mathfrak{S}$-absorbing disk is a neighborhood of 0.*

b) *In E' every subset which is bounded for the $\mathfrak{S}$-convergence is equicontinuous.*

c) *In $L(E, F)$ every subset which is bounded for the $\mathfrak{T}$-convergence is equicontinuous, for an arbitrary* LCTVS *F.*

(Compare with Proposition 3, Corollary 3.) The equivalence of a) and b) is immediate by polarity since in b) we can restrict ourselves to the weakly closed disked subsets of E', and can then apply Proposition 3, Corollary 2. b) implies c), by the characterization of bounded or equicontinuous, subsets of $L(E, F)$ by the nature of the sets $M^{-1}(V)$ (V closed disked neighborhood of 0 in F). Finally, c) implies b): it suffices to suppose c) true for a space F different from the closure of the origin, by the immersion of E' into $L(E, F)$ indicated at the beginning of Section 1.

DEFINITION 3   *Let E be an* LCTVS. *E is barrelled (resp. quasi-barrelled) if every weakly bounded (resp. strongly bounded) subset of E' is equicontinuous.*

By the preceding lemma, this means that in E, every closed absorbing disk (resp. every closed bornivorous disk) is a neighborhood of 0; furthermore:

PROPOSITION 5   *The LCTVS E is barrelled (resp. quasi-barrelled) if and only if the following proposition is true: For every locally convex space F, every subset of $L(E, F)$ bounded for the topology of pointwise convergence (resp. bounded for the topology of bounded convergence) is equicontinuous.*

COROLLARY   *Let E be a barrelled LCTVS, F a Hausdorff LCTVS, (u_i) a sequence of continuous linear mappings from E into F such that $u_i(x)$ tends to a limit $u(x)$ for every $x \in E$. Then the sequence (u_i) is equicontinuous, therefore u is a continuous linear mapping from E into F, and u_i tends to u uniformly on every compact set.*

A barrelled space is quasi-barrelled more precisely, to say that E is barrelled is to say that E is quasi-barrelled and satisfies the following supplementary condition: every weakly bounded subset of E' is strongly bounded. We have seen (Section 1, Theorem 1) that this last property is satisfied in a wide variety of cases, for example every time E is quasi-complete. Thus, *if E is quasi-complete, "barrelled" and "quasi-barrelled" have the same meaning.* We point out that a quasi-barrelled space necessarily has a Mackey topology $\tau(E, E')$, since this means that the weakly compact disks of E' are equicontinuous, and they are in any case strongly bounded.

The Banach–Steinhaus theorem (Chapter 1, Section 15) stated for *locally convex* complete metrisable spaces, can be expressed precisely as

THEOREM 2   *A metrisable and complete LCTVS is barrelled.*

It is easy to see that it is essential that the space considered be complete. We shall see in the next section that a metrisable LCTVS, even when not complete is always quasi-barrelled.

EXERCISE 1

a) The quotient space of a barrelled (resp. quasi-barrelled) space is barrelled (resp. quasi-barrelled).

b) Let (E_i) be a family of LCTVS, then $\Pi\, E_i$ is barrelled (resp. quasi-barrelled) if and only if every E_i is barrelled (resp. quasi-barrelled) (see Chapter 4, Part 1, Section 4, Exercise 7).

c) A vector subspace which is a topological direct factor of a barrelled (resp quasi-barrelled) space is barrelled (resp. quasi-barrelled).

EXERCISE 2   Let E be an LCTVS. Consider on E' a topology compatible with the duality (E, E'). Show that E is barrelled if and only if the topology of E is $\tau(E, E')$ and E' is reflexive.

EXERCISE 3   Deduce from Exercise 2 an example of a complete LCTVS with a Mackey topology and which is not barrelled (therefore not quasi-barrelled). (Take the dual E' of a metrisable and complete space E which is not reflexive and equip E' with the Mackey topology $\tau(E', E)$). Using Exercise 1 b) deduce an example of a closed not barrelled vector subspace of a complete barrelled space (recall that every Hausdorff LCTVS is isomorphic to a topological vector subspace of a product of Banach spaces).

EXERCISE 4   The completion of a quasi-barrelled space is barrelled.

EXERCISE 5   Let E be an LCTVS which is a Baire space. Show that E is barrelled (notice that the proof of the Banach–Steinhaus theorem is valid in this case). We point out that there exist barrelled normed spaces which are not Baire spaces (see Section 5, Exercise 11).

EXERCISE 5'   Let E be an LCTVS.

1) E is quasi-barrelled if and only if the canonical mapping from E into the strong dual of E'_b is an isomorphism into. It is an isomorphism onto if and only if E is quasi-barrelled and reflexive. Show that E is then barrelled.

2) Let E be a barrelled and quasi-complete LCTVS. E is reflexive if and only if E'_b is reflexive (see Chapter 2, Section 18, Theorem 13, Corollary 4).

EXERCISE 6   Show that the lemma of Chapter 1, Section 15, is true when E is a barrelled metrisable LCTVS, G a LCTVS. From this, conclude that if E and F are metrisable LCTVS and one of them is barrelled, every bilinear mapping from $E \times F$ into an LCTVS G, which is continuous with respect to each variable, is continuous; if E and F are both barrelled, every set M of separately continuous bilinear mappings from $E \times F$ into G, such that $M(x, y)$ is a bounded subset of G for every $x \in E$, $y \in F$, is equicontinuous.

EXERCISE 7   Let K be a compact space.
   a) Let H be a non-bounded set of measures on K; show that there exists $x \in K$ such that for every neighborhood V of x and every $n > 0$, there exists $\mu \in H$ and a subset A of V such that $| \mu(A) | > n$ (notice that it suffices that x be such that for every V and n we can find $\mu \in H$

such that $| \mu | (V) > n$; prove the existence of x by contradiction using Borel–Lebesgue).

b) Under these conditions either there exists a neighborhood V of x such that $| \mu | (V \cap \complement x)$ remains bounded for $\mu \in H$ (then $\mu(x)$, $\mu \in H$, does not remain bounded), or we can construct by induction a sequence of open neighborhoods V_n of x and a sequence of open sets U_n and finally a sequence (μ_n) extracted from H such that:

$$\overline{V_{n+1}} \subset V_n \cap \complement \bigcup_{1 \leqslant i \leqslant n} U_i,$$
$$U_{n+1} \subset V_{n+1} \cap \complement x$$
$$\mu_i(U_{n+1}) \leqslant 2^{-n-1} \text{ for } i \leqslant n$$
$$\mu_{n+1}(U_{n+1}) > n + 1.$$

c) From this conclude that there exists an open set U such that $\mu(U)$, $\mu \in H$, does not remain bounded (choose $U = K \cap \complement x$ for the first case in b), or choose U to be a sufficiently small neighborhood of x and $U = \bigcup U_i$ in the second case).

d) Suppose that the $\mu \in H$ are all measures of base μ_0, where μ_0 is a positive measure such that every point of K has a measure zero for μ_0. If H is not bounded, there exists an open set U, whose boundary has measure zero for μ_0, and such that $\mu(U)$, $\mu \in H$, does not remain bounded. (Show that in the construction of b) we can assume

$$\mu_0(U_n) \leqslant 1/n,$$

and furthermore that each U_n has a boundary of measure zero, using the known general result: every point a of K has a fundamental system of open neighborhoods whose boundary has measure zero. This result is independent of the hypothesis on μ_0. We conclude that the statement is still true if we replace a by an arbitrary compact subset of K.) Now consider the sequence of measures $\varepsilon_0 + n(\varepsilon_{1/n} - \varepsilon_{-1/n})$ on the compact interval $[-1, 1]$ and show that the restrictive hypothesis on μ_0 is essential: H may even be such that $\mu(U)$, $\mu \in H$, remains bounded every time U is reduced to a point, and H need not necessarily be bounded.

EXERCISE 8   Let I be a set of indices, E the vector subspace of $l^\infty(I)$ generated by the characteristic functions ϕ_A of subsets $A \subset I$.

a) Show that E (which is dense in $l^\infty(I)$) is barrelled, i.e. that every subset H of the dual of $l^\infty(I)$, bounded for the ϕ_A, is bounded. (Use the preceding Exercise, a) and b) noticing that we are considering the second case of b) and that we can suppose the U_n open and closed.

Clearly, K is the Stone compactification of I, such that $C(K)$ can be identified with $l^\infty(I)$ in the usual manner. The open and closed subsets U_i of K correspond to the subsets A_i of I, and, through a variation of the reasoning employed in Chapter 3, Section 7, Exercise 2c, reduce to the case where $I = N$ (the set of natural numbers) and where A_i is reduced to i. From this conclude that we would have $\mu_n(N) \to \infty$, which is absurd.)

b) Show that if I is infinite, the preceding barrelled space E is not a Baire space. (For every integer $n > 0$, let E_n be the subset of E formed by the

$$\sum_{i=1}^{n} \lambda_i \phi_{Ai}$$

with $|\lambda_i| \leqslant n$. Show by an argument of weak compactness that E_n is closed in $l^\infty(I)$ and a fortiori in E, then observe that E is the union of the sequence (E_n) but that every E_n has an empty interior in E.)

c) Let K be a compact Stone space, i.e. a space in which the closure of an open set is open. Let E be the vector subspace of $C(K)$ generated by the characteristic functions of simultaneously open and closed sets of K. Show that E (which is dense in $C(K)$) is barrelled, therefore every set H of measures on K such that $\mu(A)$, $\mu \in H$, is bounded for every subset A of K which is simultaneously open and closed. Reduce to a) with the aid of the additional known fact: there exists a projection of norm 1 of $l^\infty(I)$ on $C(K)$, which is a homomorphism for the algebraic structures and therefore transforms idempotents, i.e. characteristic function of sets, into idempotents.)

4 Bornological spaces

DEFINITION 4　*Let E be an LCTVS E is bornological if every bornivorous disk of E is a neighborhood of 0.*

A fortiori a closed absorbing disk is a neighborhood of 0, thus a *bornological space is quasi-barrelled*, a fortiori its topology is the Mackey topology. The following proposition expresses the interest which arises from the concept of bornological space.

PROPOSITION 6　*Let E be an LCTVS. The following conditions are equivalent:*

　a) *E is bornological;*

　b) *Every set M of linear mappings from E into an LCTVS F such that*

$M(A)$ *is a bounded subset of F for every bounded subset A of E, is equi-continuous;*

c) *Every linear mapping u from E into an* LCTVS *F which takes bounded sets into bounded sets, is continuous;*

d) *Every set of linear forms on E, uniformly bounded on the bounded subsets of E, is equicontinuous;*

e) *Every linear form on E, bounded on the bounded sets, is continuous, and the topology of E is the Mackey topology* $\tau(E, E')$.

a) implies b) since the hypothesis on M means that for every disked neighborhood V of 0 in F, $M^{-1}(V)$ is a bornivorous disk in E (Section 2, Proposition 3), therefore it is a neighborhood of 0 since E is bornological, whence M is equicontinuous; b) implies c) trivially. We shall show that c) implies a). Let, in fact, V be a bornivorous disk in E, let F be the space E with the semi-norm gauge of V, we wish to show that V is a neighborhood of 0, i.e. that the identity mapping of E onto F is continuous. But to say that V is bornivorous is to say that the identity mapping from E onto F transforms bounded sets into bounded sets (Proposition 3), therefore it is continuous by hypothesis. Thus a), b), c) are equivalent and these imply d) which in turn implies e) (since every weakly compact disk of E' will be equicontinuous). Finally, e) implies c) since the topology of E is $\tau(E, E')$, in order to verify the continuity of u it suffices to verify that it is weakly continuous (Chapter 2, Section 16, Proposition 28, Corollary 2), therefore for every $y' \in F'$, $y' \circ u$ is a continuous linear form on E. Now this linear form is in fact bounded on the bounded subsets of E.

Bornological spaces are of value as they allow the criterion given in Proposition 6c to be used in deciding whether or not a linear mapping u from the bornological space E into the LCTVS F is continuous. In fact, it is not difficult to verify whether or not u transforms the bounded subsets of E into bounded subsets of F (using the closed graph theorem, when the spaces E_A—where A is a closed and bounded disk in E—are complete). We now give an easily applicable characterization of linear mappings from one LCTVS E into another, which mappings transform bounded subsets into bounded subsets. We need

DEFINITION 5 *Let E be an* LCTVS *and* (x_i) *a sequence in E. We say that* (x_i) *tends to a limit* $x \in E$ *in the sense of Mackey if there exists a bounded disk A in E such that* (x_i) *tends to x in the normed space* E_A.

This means that $x_i - x$ tends to zero in the sense of Mackey; to say that x_i tends to 0 in the sense of Mackey is to say that there exists a sequence of scalars $\lambda_i > 0$ tending to zero, such that the sequence of x_i/λ_i remains bounded; if we replace the sequence (λ_i) by $(\sqrt{\lambda_i})$ we see that we can even suppose that x_i/λ_i tends to zero. It is clear that convergence in the sense of Mackey implies convergence in the sense of the topology of E.

PROPOSITION 7   *Let u be a linear mapping from one LCTVS E into another, F. The following conditions are equivalent:*

a) *u transforms bounded subsets of E into bounded subsets of F.*

b) *u transforms the sequences which converge to 0 in the sense of Mackey into sequences which converge to 0 in the sense of Mackey.*

c) *u transforms the sequences which converge to 0 in the sense of Mackey into sequences which converge to 0 (for the topology of F).*

d) *u transforms the sequences which converge to 0 in the sense of Mackey into bounded sequences.*

That a $\Rightarrow$ b $\Rightarrow$ c $\Rightarrow$ d is immediate. In order to show that d) $\Rightarrow$ a), we suppose that a) is not true and therefore there exists a bounded subset A of E and a neighborhood V of 0 in F such that $u(A)$ is not contained in any homothetic of V. Then for every integer $n > 0$, an $x_n \in A$ exists such that $u(x_n) \notin n^2 V$, from which we have $u(x_n/n) \notin nV$, therefore x_n/n would be a sequence in E converging to zero in the sense of Mackey whose image in F is not bounded. Hence, by contradiction, the required result. From Proposition 7 we find new statements equivalent to condition b) of Proposition 6. We obtain for example:

COROLLARY   *Let E be a bornological LCTVS, u a linear mapping from E into an LCTVS F. The mapping u is continuous if and only if u is continuous for the sequences; or equivalently, u transforms the sequences which converge to zero into bounded sequences; it is sufficient for this to be true of the sequences which converge to zero in the sense of Mackey.*

We now give the following

LEMMA   *Let E be a metrisable LCTVS, (x_i) a sequence in E converging to zero; then it converges to zero in the sense of Mackey.*

In fact, let (p_n) be a fundamental sequence of continuous seminorms of E; the hypothesis then means that for every n, the sequence $(p_n(x_i))$ tends to zero. We must conclude that there exists a sequence of scalars $\lambda_i > 0$ tending to zero such that x_i/λ_i remains bounded, therefore such that for every n, the sequence $(p_n(x_i)/\lambda_i)$ remains bounded. For this it suffices to set

$$M_n = \sup p_n(x_i)$$

and choose

$$\lambda_i = \Sigma \, \frac{1}{2^n M_n} \, p_n(x_i).$$

Using the lemma and Criterion c) of Proposition 7, we see that the bounded linear mappings from E into an LCTVS F are exactly those which are continuous for the sequences, i.e. (E metrisable) those which are continuous. Then we have (Proposition 6c)

THEOREM 3   *A metrisable LCTVS (not necessarily complete) is bornological.*

COROLLARY   *A metrisable LCTVS is quasi-barrelled and a fortiori its topology is the Mackey topology.*

From this last fact, we deduce (see Chapter 2, Section 16, Proposition 28, Corollary 2) the following corollaries:

COROLLARY 1   *Let u be a linear mapping from a metrisable LCTVS E into an LCTVS F. The mapping u is continuous if and only if u is weakly continuous.*

COROLLARY 2   *Let u be a continuous linear mapping from an LCTVS F into an LCTVS E. The mapping u is a homomorphism if and only if u is a weak homomorphism.*

EXERCISE 1
   a) A quotient space of a bornological space is bornological.

   b) Let (E_i) be a finite family of LCTVS. Their product is bornological if and only if the E_i are bornological.

   c) A topological direct factor of a bornological space is bornological.

EXERCISE 2   Let $(E_i)_{i \in I}$ be a family of LCTVS. For the product $\prod_{i \in I} E_i$ to be bornological it is sufficient, and it is necessary if the E_i are Hausdorff and $\neq 0$, that the E_i be bornological and that the product R^I be bornological. From this, conclude that the product of a sequence of bornological spaces is bornological.

EXERCISE 3   Let E be an LCTVS. Show that there exists on E a topology T which is the coarsest of those bornological topologies which are finer than the given topology; the bornivorous disks of E form a fundamental system of neighborhoods of 0 for T.

EXERCISE 4   Show that the results of Section 2, Exercise 3, remain valid if we replace the Banach space E, or H, by a quasi-complete bornological space.

EXERCISE 5   Let E be a Hausdorff LCTVS. The following conditions are equivalent:

a) For every closed and bounded disk A of E, E_A is complete.

b) For every bounded disk A of E, there exists a bounded disk B in E containing A such that E_B is complete.

c) For every continuous linear mapping u from a normed space H in E, there exists a continuous linear mapping from the completion of H into E which extends u.

d) The closed convex hull of every sequence in E, which converges to zero in the sense of Mackey, is compact. (For this, use Chapter 2, Section 3, Exercise 3.)

Then every bounded subset of E is strongly bounded, whence every set of continuous linear mappings from E into a LCTVS F, which is bounded for pointwise convergence, is bounded for bounded convergence.

EXERCISE 6   Let E be a metrisable LCTVS. The following conditions are equivalent:

a) E is barrelled;

b) Every weakly bounded subset of E' is strongly bounded;

c) Every sequence weakly convergent to 0 in E' converges uniformly on every compact set;

e) E is not the union of a sequence of closed *disks* none of which is a neighborhood of 0.

EXERCISE 7   Let E be a bornological (resp. quasi-barrelled) space, F a complete (resp. quasi-complete) LCTVS. Show that $L_b(E, F)$ is complete (resp. quasi-complete).

5 Bilinear functions: Types of continuity. Continuity and separate continuity

DEFINITION 6   *Let E, F, G be topological spaces, u a mapping from $E \times F$ into G. The mapping u is separately continuous if u is continuous with respect to each variable. If G is furthermore a uniform space, we say that set M of mappings $u(x, y)$ from $E \times F$ into G is equicontinuous in x*

(resp. in y) if for every $y \in F$, the set of mappings $x \mapsto u(x, y)$ where u runs through M, is an equicontinuous set of mappings from E into G (resp. if . . .). M is separately equicontinuous if it is equicontinuous both in x and in y.

Interpreting the mappings of $E \times F$ into G as mappings from E into the set of mappings of F into G, to say that u is separately continuous is to say that there corresponds a mapping from E into the space $C(F, G)$ of continuous mappings from F into G, which is continuous when $C(F, G)$ has the topology of pointwise convergence. Similarly, to say that M is a set of separately continuous mappings from $E \times F$ into G which is *equicontinuous with respect to x* is to say that the set of mappings from E into the space $C_s(F, G)$ which corresponds to it, is equicontinuous; to say that M is *equicontinuous with respect to y* is to say that for every $x \in E$ the set $M(x)$ of images of x by the mappings $E \to C_s(F, G)$ corresponding to $u \in M$ is an equicontinuous subset of $C(F, G)$. Also, if E, F, G are LCTVS, to say that a mapping of $E \times F$ into G is bilinear is to say that the mapping of E into the space of linear mappings from F into G corresponding to it, maps E into the space of linear mappings from F into G and that it is linear. Then we have the following trivial but fundamental fact:

THEOREM 4 *Let E, F, G be LCTVS. Then the separately continuous bilinear mappings from $E \times F$ into G are in bijective correspondence with the continuous linear mappings from E into the space $L_s(F, G)$ of continuous linear mappings from F into G, equipped with the topology of pointwise convergence.*

COROLLARY *Let E and F be LCTVS. Then the separately continuous bilinear forms on $E \times F$ correspond bijectively to the continuous linear mappings from E into the weak dual F'_s of F.*

Notice that the notion of a bilinear *form* separately continuous on $E \times F$ depends only on the duals of E and F. It is clear that we can obtain analogous statements by exchanging the roles of E and F in the above. Then,

PROPOSITION 8 *Let E, F, G be LCTVS and u a bilinear mapping from $E \times F$ into G. The mapping u is continuous if and only if it is continuous at the origin. If G is normable it is also necessary and sufficient that the corresponding mapping from E into $L(F, G)$ transforms some neighborhood of 0 in E into an equicontinuous subset of $L(F, G)$. There is an analogous statement for the equicontinuous sets of bilinear mappings from $E \times F$ into G.*

The proof is immediate. In particular:

COROLLARY 1   *Let E, F be LCTVS. Then the continuous bilinear forms on $E \times F$ correspond bijectively to the linear mappings from E into F' that transform some neighborhood of 0 into an equicontinuous subset of F'. The equicontinuous sets of bilinear forms correspond to the sets of linear mappings from E into F' which map a certain neighborhood of 0 into a fixed equicontinuous subset of F'.*

COROLLARY 2   *Let E be an LCTVS and let E' have the strong topology. Then the canonical bilinear form on $E \times E'$ is continuous if and only if E is semi-normable.*

In fact, by Corollary 1, this bilinear form is continuous if and only if there exists a neighborhood of 0 in E which is equicontinuous as a subset of the dual of E', i.e. which is bounded. Now this means that E is semi-normable (Chapter 1, Section 7, Proposition 12, Corollary 1). Corollary 2 shows that there exist very important bilinear forms which are not continuous but only separately continuous. Recall however that if E and F are metrisable and complete, then every separately continuous bilinear mapping on $E \times F$ is continuous (Chapter 1, Section 15, Theorem 12). We shall see an analogous important case in Chapter 5, Part 3, Section 2.

Hypocontinuity   We therefore wish to find types of continuity of bilinear forms which are weaker than continuity but stronger than separate continuity.

LEMMA   *Let E, F be topological spaces, G a uniform space, $\mathfrak{G}$ a set of subsets of E, u a separately continuous mapping from $E \times F$ into G. Then the following conditions are equivalent:*

a) *the mapping $E \to C_s(F, G)$ corresponding to u transforms the $A \in \mathfrak{G}$ into equicontinuous subsets of $C(F, G)$;*

b) *the mapping $F \to C_s(E, G)$ corresponding to u is also a continuous mapping from F into the space $C_{\mathfrak{G}}(E, G)$ of continuous mappings from E into G with the $\mathfrak{G}$-topology.*

When these conditions are satisfied, for every $A \in \mathfrak{G}$, the restriction of u to $A \times F$ is continuous. When, furthermore, the analogous conditions relative to a set $\mathfrak{T}$ of subsets of F are satisfied, for every $A \in \mathfrak{G}$, $B \in \mathfrak{T}$, the restriction of u to $A \times B$ is uniformly continuous.

The proof is trivial (see Chapter 0, Proposition 9, 3). The lemma can be generalized to refer to a *set* of separately continuous mappings from

$E \times F$ into G, and a further variant of the statement concerns the case in which we suppose that F is also a uniform space, and thus replace in conditions a) and b), "continuous" by "uniformly continuous" and "equicontinuous" by "uniformly equicontinuous".

DEFINITION 6'   *Let E, F be topological spaces, G a uniform space, $\mathfrak{S}$ a set of subsets of E. A mapping from $E \times F$ into G is hypocontinuous with respect to $\mathfrak{S}$, if it is separately continuous and satisfies the conditions of the preceding lemma. Likewise we define for a set of mappings from $E \times F$ into G the notion of equihypocontinuity of M with respect to $\mathfrak{S}$. Also, if F is a uniform space we have the following variants: the mapping u uniformly hypocontinuous with respect to $\mathfrak{S}$, the set M of mappings uniformly equihypocontinuous with respect to $\mathfrak{S}$.*

When E, F, G are LCTVS and we consider only bilinear mappings from $E \times F$ into G, since the equicontinuous subsets of $L(F, G)$ are already uniformly equicontinuous, there is no need to distinguish between a hypocontinuous or uniformly hypocontinuous bilinear mapping nor between a set of equihypocontinuous or uniformly equihypocontinuous bilinear mappings (with respect to a given $\mathfrak{S}$). In this case, the notion is of interest only if $\mathfrak{S}$ is a set of bounded subsets of E. Clearly, the preceding paragraph could be repeated exchanging the roles of E and F; thus, if $\mathfrak{T}$ is a set of subsets of F, we have the concept of a mapping of $E \times F$ hypocontinuous with respect to $\mathfrak{T}$, etc. If we have simultaneously a set $\mathfrak{S}$ of subsets of E and a set $\mathfrak{T}$ of subsets of F, we obtain the concept of a hypocontinuous mapping with respect to $\mathfrak{S}$ and $\mathfrak{T}$, etc. We recall the meaning of the concepts introduced where considering bilinear mappings:

PROPOSITION 9   *Let E, F, G be LCTVS, $\mathfrak{S}$ a set of bounded subsets of E. A separately continuous bilinear mapping u from $E \times F$ into G is hypocontinuous with respect to $\mathfrak{S}$ if and only u satisfies one of the following equivalent condition:*

a) *the mapping $E \to L(F, G)$ corresponding to u transform the $A \in \mathfrak{S}$ into equicontinuous subsets of $L(F, G)$.*

b) *the mapping $F \to L(E, G)$ corresponding to u is continuous when we equip $L(E, G)$ with the topology of the $\mathfrak{S}$-convergence.*

Then for every $A \in \mathfrak{S}$ the restriction of u to $A \times F$ is continuous. There is an analogous statement for a set of separately continuous bilinear mappings which is equihypocontinuous with respect to $\mathfrak{S}$.

COROLLARY 1   *Let E, F, G be LCTVS, u a separately continuous bilinear* *mapping from* $E \times F$ *into* G *(resp.* M *a pointwise bounded set of continuous bilinear mappings from* $E \times F$ *into* G*). If* F *is barrelled,* u *is hypocontinuous with respect to the bounded subsets of* E *(resp.* M *is equihypocontinuous with respect to the bounded subsets of* E*).*

Proof This follows from criterion a) of Proposition 9 and from the fact that in $L_s(F, G)$ every bounded subset is equicontinuous.

A bilinear mapping from $E \times F$ into G is *hypocontinuous*, if it is hypocontinuous with respect to the *bounded* subsets of E and F; likewise for an *equihypocontinuous* set of linear mappings from $E \times F$ into G. Notice that for a *form* or a set of bilinear forms separately continuous on $E \times F$, hypocontinuity with respect to a set $\mathfrak{S}$ of subsets of E does not depend on the topology of E but only on the dual of E, so that we can choose the weak topology on E.

COROLLARY 2 *Let* u *be a hypocontinuous bilinear form on* $E \times F$*. Then for every bounded subset* A *of* E *and every bounded subset* B *in* F*, the restriction of* u *to* $(A_{\mathrm{weak}}) \times F$ *and to* $E \times (B_{\mathrm{weak}})$ *is continuous. Likewise for the restriction of* u *to* $(A_{\mathrm{weak}}) \times (B_{\mathrm{weak}})$ *if either* A *or* B *is precompact.*

The last assertion follows from the fact that on a precompact subset of an LCTVS, the induced topology is identical to the induced *weak* topology (Chapter 2, Section 18, Proposition 33).

Normally in analysis we can show by standard techniques that the separately continuous bilinear mappings we find have the hypocontinuity properties we need. For example, see Corollary 1 of Proposition 9 and Exercises 7, 8, 9 which follow. The value of hypocontinuity is seen in the obvious remark that if u is a separately continuous bilinear mapping from $E \times F$ into G, hypocontinuous with respect to the compact sets of E (or of F), then u is continuous with respect to convergent sequences; continuity with respect to convergent sequences of $E \times F$ suffices in many cases as a substitute for continuity (see Exercise 6). On the other hand, the hypocontinuity of a bilinear mapping is useful in matters of extensions.

Extensions of hypocontinuous bilinear mappings

PROPOSITION 10 *Let* E, F, G *be LCTVS,* E_1 *(resp.* F_1*) a dense vector subspace of* E *(resp.* F*),* $\mathfrak{S}$ *(resp.* $\mathfrak{T}$*) a directed set of bounded subsets of* E_1 *(resp.* F_1*) that generates the vector space* E_1 *(resp.* F_1*),* $\overline{\mathfrak{S}}$ *(resp.* $\overline{\mathfrak{T}}$*) the set of closures in* E *(resp.* F*) of elements of* $\mathfrak{S}$ *(resp.* $\mathfrak{T}$*). We suppose that* $\overline{\mathfrak{S}}$ *generates* E*, that* $\mathfrak{T}$ *generates* F*, that* G *is Hausdorff and quasi-complete. Then every bilinear mapping* u *from* $E_1 \times F_1$ *into* G*, hypocontinuous with* *respect to $\mathfrak{S}$ and $\mathfrak{T}$, can be extended in a unique way into a bilinear mapping $\tilde{u}$ from $E \times F$ into G, hypocontinuous with respect to $\overline{\mathfrak{S}}$ and $\mathfrak{T}$. If u is continuous, $\tilde{u}$ is continuous; if u runs through a set of bilinear mappings, equihypocontinuous with respect to $\overline{\mathfrak{S}}$ and $\mathfrak{T}$, then $\tilde{u}$ runs through a set of bilinear mappings, equihypocontinuous with respect to $\overline{\mathfrak{S}}$ and $\mathfrak{T}$.*

The proof is immediate and left to the reader, likewise for the following variant of Proposition 10:

PROPOSITION 11 *Let E, F be Hausdorff LCTVS. Then every hypocontinuous bilinear form on $E \times F$ can be uniquely extended to a separately weakly continuous bilinear form on $E \times F''$. When u runs through an equicontinuous set of bilinear forms on $E \times F$, then $\tilde{u}$ runs through an equicontinuous set of bilinear forms on $E \times F''$ equihypocontinuous with respect to the bounded subsets of E and the equicontinuous subsets of F'' (considered as duals of E'_b and F'_b). When u runs through an equicontinuous set of bilinear forms on $E \times F$, then $\tilde{u}$ runs through an equicontinuous set of bilinear forms on $E \times F''$.*

(Recall that unless otherwise stated, the weak topology of the bidual E'' means the topology $\sigma(E'', E')$, while the "natural topology" of E^{11}, that we always consider when there is no other topology of E'', is the topology of uniform convergence on the equicontinuous subsets of E'.)

EXERCISE 1 Let E, F, G be LCTVS, M a set of bilinear mappings from $E \times F$ into G. The set M is equicontinuous (or equihypocontinuous with respect to a set $\mathfrak{S}$ of bounded subsets of E or F; or separately equicontinuous) if and only if for every equicontinuous subset C' of G', the set $M_{C'}$ of bilinear forms $\langle u(x, y), z' \rangle$ on $E \times F$ where $u \in M$ and $z' \in C'$ has the same property. (Identify G with a space of functions on G' with the topology of uniform convergence on the set $\mathfrak{T}$ of equicontinuous subsets; the problem is then reduced to a general topological situation independent of vector structures.)

EXERCISE 2 Let E, F, G be LCTVS, $\mathfrak{S}$ a set of bounded disks of E, M a set of bilinear mappings separately continuous from $E \times F$ into G. For the set M to be equihypocontinuous with respect to $\mathfrak{S}$, it suffices already (and it is necessary) that for every $A \in \mathfrak{S}$, the set of restrictions of $u \in M$ to $A \times F$ be equicontinuous at $(0, 0)$.

EXERCISE 3 *Let E, F, G be LCTVS, u a bilinear mapping from $E \times F$ into G, hypocontinuous with respect to the compact subsets of E (or of F). Let T be a metrisable topological space, f a continuous mapping*

I

from T into E, g a continuous mapping from T into F. Show that the mapping $t \mapsto u(f(t), g(t))$ from T into G is continuous.

EXERCISE 4   Let E, F, G be LCTVS, M a set of separately continuous linear mappings from $E \times F$ into G. Show that if M is equicontinuous with respect to *one* of the variables, then $M(x, y)$ is a bounded subset of G for every

$$(x, y) \in E \times F.$$

Let A (resp. B) be a bounded subset of E (resp. F). Show that $M(A \times B)$ is a bounded subset of G in each of the following cases:

a) $M(x, y)$ is a bounded subset of G for every $(x, y) \in E \times F$, and A and B are strongly bounded.

b) M is equicontinuous with respect to x and A is *strongly bounded*;

c) M is equihypocontinuous with respect to the compact subsets of E (or even only with respect to the sequences in E tending to 0 in the sense of Mackey).

EXERCISE 5   Show that in the preceding exercise, it is not sufficient in a) to suppose A or B strongly bounded nor in b) to suppose that B is strongly bounded (A supposed only bounded), while in a) and b) the hypothesis on M remains the same. Finally, in c) it is not sufficient to suppose M separately equicontinuous. (We shall show that there are spaces where there exist bounded subsets not strongly bounded; then one can put in a) and b) $F = G = k = $ field of scalars, and in c) $G = k, F = E'_s$, M reduced to one element: the canonical bilinear form on $E \times E'$.) From this deduce counter-examples in the case where we suppose M reduced to a point (establish a converse of Exercise 1).

EXERCISE 6   Let E, F, G be LCTVS, u a bilinear mapping from $E \times F$ into G such that u $(A \times B)$ is a bounded subset of G for every bounded subset A of E and every bounded subset B of F. Let U be an open subset of R^n, let f (resp. g) be a continuously differentiable mapping from U into E (resp. F). Show that the function $t \mapsto u(f(t), g(t))$ on U with values in G is continuously differentiable and that its derivative with respect to t_i is given by

$$\frac{\partial}{\partial t_i} u(f(t), g(t)) = u\left(\frac{\partial}{\partial t_i} f(t), g(t)\right) + u\left(f(t), \frac{\partial}{\partial t_i} g(t)\right).$$

(Show first of all that a derivative given by the preceding formula exists; then show that if u is as given above and f' a continuously differentiable mapping from U into E, g a continuous mapping from U

into G, then the mapping $t \mapsto u(f'(t), g(t))$ from U into G is continuous). Establish a quicker proof in the case where we suppose u hypocontinuous with respect to the compact subsets of E (or of F).

EXERCISE 7   Let E, F, G be LCTVS, M a separately equicontinuous set of bilinear mappings from $E \times F$ into G. Show that M is equi-hypocontinuous with respect to the compact subsets of F, when we equip E with the topology of uniform convergence on the weakly compact subsets of E' (case where G is the field of scalars with the aid of Exercise 1, then interpret M as an equicontinuous set of linear mappings from F into E' weak; notice that M is relatively compact in $L(F, E')$ for compact convergence by Ascoli's theorem; from this conclude using continuity that for every compact $K \subset F$, $M(K)$ is a weakly relatively compact subset of E').

EXERCISE 8   Let E, F, G be LCTVS, M a set of separately continuous bilinear mappings from $E \times F$ into G, $\mathfrak{T}$ a set of strongly bounded subsets of F. We suppose F to be barrelled (or quasi-barrelled) and that for every $(x, y) \in E \times F$, $M(x, y)$ is a bounded subset of G (or that M is equicontinuous with respect to E). Show that M is equihypocontinuous with respect to $\mathfrak{T}$.

EXERCISE 9   Let E, F, G be LCTVS, $\mathfrak{G}_1$ (resp. $\mathfrak{G}_2$) a set of bounded subsets of E (resp. F) covering E (resp. F). We suppose that for every continuous linear mapping u from E into F we have $u(\mathfrak{G}_1) \subset \mathfrak{G}_2$. Then the bilinear mapping $(v, u) \mapsto v - u$ from $L_{\mathfrak{G}_2}(F, G) \times \circ L_{\mathfrak{G}_1}(E, F)$ into $L_{\mathfrak{G}_1}(E, G)$ is separately continuous. Let Σ be the set of subsets M of $L_{\mathfrak{G}_1}(E, F)$ such that for every $A \in \mathfrak{G}_1$, $M(A)$ is a subset of F belonging to $\mathfrak{G}_2$. Then $(v, u) \mapsto v \circ u$ is hypocontinuous with respect to Σ and with respect to the set of equicontinuous subsets of $L_{\mathfrak{G}_2}(F, G)$. Particular cases are:

a) If $\mathfrak{G}_2$ is the set of all bounded subsets of F, then $(v, u) \mapsto v \circ u$ is hypocontinuous with respect to the set of bounded subsets of $L_{\mathfrak{G}_1}(E, F)$ and the set of equicontinuous subsets of $L_{\mathfrak{G}_2}(F, G)$.

b) If $\mathfrak{G}_1$ (resp. $\mathfrak{G}_2$) is the set of finite subsets of E (resp. F) then

$$(v, u) \mapsto v \circ u$$

is a hypocontinuous mapping with respect to the equicontinuous subsets of $L_{\mathfrak{G}_2}(F, G)$.

c) If $\mathfrak{G}_1$ is a set of compact subsets of E, $\mathfrak{G}_2$ the set of all compact subsets of F, then $(v, u) \mapsto v \circ u$ is a hypocontinuous mapping with respect to the compact subsets of $L_{\mathfrak{G}_1}(E, F)$ and with respect to the equicontinuous subsets of $L_{\mathfrak{G}_2}(F, G)$.

EXERCISE 10   Let F be a Banach space, let $E = F'$ be the dual Banach space, u the canonical bilinear form on $E \times F$. Let v_1 be the canonical extension by weak continuity of u to $E \times F''$, and w_1 the analogous extension of v_1 to $E'' \times F''$. We define likewise the extension v_2 of u to $E'' \times F$, and the extension w_2 from v_2 to $E'' \times F''$. Make w_1 and w_2 explicit (which are bilinear forms on $E'' \times F''$), and show that they are distinct and not weakly separately continuous if F is not reflexive. Show then that there does not exist any bilinear form weakly separately continuous on $E'' \times F''$ which extends u.

6 Spaces of bilinear mappings. Definitions and notations

Let E, F, G be LCTVS, let $\mathfrak{G}$ be a set of bounded subsets of E, $\mathfrak{T}$ a set of bounded subsets of F, H a space of bilinear mappings separately continuous from $E \times F$ into G. Consider on H a $\mathfrak{G} \times \mathfrak{T}$ topology, where $\mathfrak{G} \times \mathfrak{T}$ stands for the set of subsets of $E \times F$ of type $A \times B$ with $A \in \mathfrak{G}$, $B \in \mathfrak{T}$. This topology is locally convex if and only if for every $A \times B \in \mathfrak{G} \times \mathfrak{T}$ and every $u \in H$, the set $u(A \times B)$ is a bounded subset of G (Chapter 1, Section 8, Theorem 3).

PROPOSITION 12   *On the space of bilinear mappings H the $\mathfrak{G} \times \mathfrak{T}$-topology is identical to the topology induced by the space of linear mappings from E into $L_\tau(F, G)$ with $\mathfrak{G}$-convergence.*

The verification is immediate, see Chapter 0, Proposition 6''. In particular, if the $u \in H$ are hypocontinuous with respect to $\mathfrak{G}$ or $\mathfrak{T}$, then the corresponding mappings from E into $L_\mathfrak{T}(F, G)$ transform the $A \in \mathfrak{G}$ into bounded subsets of $L_\mathfrak{T}(F, G)$ and therefore H is locally convex. Likewise, applying Proposition 3', (Section 2), we see that H is always locally convex if the $A \in \mathfrak{G}$ or the $B \in \mathfrak{T}$ are *strongly* bounded. In particular, by the Banach–Steinhaus–Mackey theorem (Section 2), we find the

COROLLARY   *If the elements of $\mathfrak{G}$ or $\mathfrak{T}$ are complete disks, then the space of all separately continuous bilinear mappings from $E \times F$ into G with the $\mathfrak{G} \times \mathfrak{T}$-topology is locally convex.*

The $\mathfrak{G} \times \mathfrak{T}$-topology is the *topology of bounded (resp. compact, resp. simple)* if $\mathfrak{G}$ and $\mathfrak{T}$ are the sets of bounded *convergence* (resp. compact, resp. finite) subsets of E and F. The space of separately continuous (resp. continuous) bilinear mappings from $E \times F$ into G is denoted by $\mathfrak{B}(E, F; G)$ (resp. $B(E, F; G)$), and we omit G when it is the field of scalars, therefore $\mathfrak{B}(E, F)$ and $B(E, F)$ are respectively the space of separately continuous bilinear forms and the space of continuous bilinear forms of $E \times F$. If we consider the topology of bounded (resp. compact, resp. pointwise) convergence on one of the preceding spaces, we indicate the fact with the index b (resp. c, resp. s) as in $\mathfrak{B}_b(E, F; G)$, $\mathfrak{B}_c(E, F; G)$, $\mathfrak{B}_s(E, F; G)$. The space $\mathfrak{B}_s(E, F; G)$ and the space $B_b(E, F; G)$ are always locally convex. When we consider the strong duals E' and F' of the LCTVS E and F we may wish to equip the space $\mathfrak{B}(E', F'; G)$ and its subspaces with the topology of uniform convergence on the products $A \times B$, where A (resp. b) is an equicontinuous subset of E' (resp. F'). This topology, always locally convex (the weakly closed equicontinuous subsets, therefore weakly compact in the dual, are strongly complete by Lemma 2 of Section 2), is called the *topology of bi-equicontinuous convergence*. We indicate this topology with the index e. When F and G are the field of scalars, we find E'' with the usual topology.

Case of normed spaces The following is a generalization of Chapter 1, Section 5, Theorem 2:

PROPOSITION *Let E_i $(i = 1, 2, \ldots, n)$ and F be semi-normed spaces, let u be an n-linear mapping from ΠE_i into F. The mapping u is continuous if and only if $\| u \|$ defined by*

$$\| u \| = \sup_{(xi), \| xi \| \leqslant |} \| u(x_1, \ldots, x_n) \|$$

$$\| x_i \| \leqslant 1$$

bi-equicontinuous is finite.

Then we have,

$$\| u(x_1, \ldots, x_n) \| \leqslant \| u \| \| x_1 \| \ldots \| x_n \|$$

for every $(x_i) \in \Pi E_i$. The expression $\| u \|$ on $B(E_1, \ldots, E_n; F)$ is a semi-norm (it is a norm if F is Hausdorff), the corresponding topology being the topology of uniform convergence on the product of the unit balls of E_i (or on the product of bounded sets). This space is complete when F is Hausdorff and complete. Finally, we have a canonical metric isomorphism from

$$B(E_1, \ldots, E_p; B(E_{p+1}, \ldots, E_n; F)) \quad \text{onto} \quad B(E_1, \ldots, E_n; F).$$

In particular, if u is a continuous bilinear form on the product $E \times F$ of two semi-normed spaces, its norm, as a bilinear form, is identical to the norm of the corresponding linear mapping from E into F' (or from F into E'). The canonical extension of the form u on $E \times F$ into a form on $E \times F''$ (see Proposition 11) has the same norm as u. Notice that if E, F, G are semi-normed spaces, then on $\mathfrak{B}(E', F'; G)$ the topology of bi-equicontinuous convergence is none other than that defined by the natural semi-norm of this space.

The spaces $E \hat{\otimes} F$ Let E, F be Hausdorff LCTVS; consider $E \otimes F$ as space of bilinear forms on $E' \times F'$; to $x \otimes y$ there corresponds the bilinear form $\langle x, x' \rangle \langle y, y' \rangle$. Since this last form is clearly weakly separately continuous (and even weakly continuous), we can consider $E \otimes F$ as a vector subspace of $\mathfrak{B}(E'_s, F'_s)$, and equip it with the topology of bi-equicontinuous convergence, for which it is a topological vector subspace of $\mathfrak{B}_e(E'_s, F'_s)$.

DEFINITION 7.   *Let E, F be Hausdorff LCTVS. We denote by $E \hat{\otimes} F$ the completion of $E \otimes F$ equipped with the topology of bi-equicontinuous convergence (the topology induced by $\mathfrak{B}_e(E'_s, F'_s)$).*

If $\mathfrak{B}_e(E'_s, F'_s) \approx L_e(E', F)$ is complete, i.e. if E and F are complete (Section 1, Proposition 1) then $E \hat{\otimes} F$ is the closed topological vector subspace of $\mathfrak{B}_e(E'_s, F'_s)$, the closure of $E \otimes F$. On the other hand let $u \in \mathfrak{B}_e(E'_s, F'_s)$; to say that the linear mapping from E' into F, which corresponds to it, transforms the equicontinuous subsets into precompact subsets of F, or into relatively compact subsets of F, is to say that the linear mapping from F' into E which corresponds to it has the analogous property (Chapter 2, Section 18, Theorem 12, Corollary); on the other hand the subspace K of

$$\mathfrak{B}_e(E'_s, F'_s) \approx L_e(E', F)$$

formed by those u is *closed* (Chapter 0, Section 4, Proposition 6'). Therefore, if E and F are complete, $E \hat{\otimes} F$ can also be identified with a vector subspace of K, in other words, the linear map from E' into F corresponding to a $u \in E \hat{\otimes} F$ transforms the equicontinuous subsets of E' into relatively compact subsets of F. The converse is true in all known cases; that is, in all known cases, every weakly continuous linear mapping from E' into F which transforms the equicontinuous subsets into relatively compact ones is a uniform limit, on the equicontinuous subsets of E, of weakly continuous linear mapping of finite rank (see Exercise 4).

If E and F are normed spaces, then the topology of the bi-equicontinuous convergence on $E \otimes F$ is defined by the natural norm of the space $B(E', F')$ of continuous bilinear forms on $E' \times F'$. We can then consider $E \hat{\otimes} F$ as a Banach space, a normed subspace of $B(E', F')$.

EXERCISE 1   Let E, F be Banach spaces; consider the space $B(E, F)$ as a subspace of the dual of $E \otimes F$.

a) Show that the unit ball of $B(E, F)$ is weakly compact, and is therefore the unit ball of the dual of $E \otimes F$ for some uniquely determined norm on $E \otimes F$.

b) $E \otimes F$ being thus normed, show that a bilinear mapping u from $E \otimes F$ into a Banach space G is continuous if and only if the linear mapping $\tilde{u}$ from $E \times F$ into G corresponding to it is continuous, then, the norms of u and $\tilde{u}$ are identical.

c) Show that $E \otimes F$ is barrelled (notice that this assertion is equivalent to Chapter 1, Section 15, Theorem 12).

d) Let P_n be the set of elements of $E \otimes F$ of rank $\leqslant n$. Show that P_n is a closed subset of $E \otimes F$ (we shall show that P_n is even closed for $\sigma(E \otimes F, E' \otimes F')$, using the following exercise). From this conclude that if E and F are finite dimensional, then $E \otimes F$ is a normed barrelled space which is *meager* (therefore not a Baine space).

EXERCISE 2   Let E, F be vector spaces, u a bilinear form on $E \times F$. We call the rank of the linear mapping from E into the algebraic dual F^* of F defined by u the rank of u. (This is clearly identical to the rank of the linear mapping from F into E^* defined by u.) Show that the rank of u is $\leqslant n$, (a given integer > 0), if and only if for every sequence of n elements (x_i) in E and every sequence of n elements (y_i) in F we have $\det (u(x_i, y_i)) = 0$. From this conclude that the set of bilinear forms on $E \times F$ of rank $\leqslant n$ is closed for the topology of pointwise convergence.

EXERCISE 3
  a) Let (E_i), (F_i) be two families of Hausdorff LCVTS, let $E = \Pi E_i$, $F = \Pi F_i$. Show that $E \hat{\otimes} F$ is canonically isomorphic to

$$\prod E_i \hat{\otimes} F_j.$$

b) Let E_i, F_i be Hausdorff LCTVS ($i = 1, 2$), let u_i be a continuous linear mapping from E_i into F_i; show that the mapping $u_1 \otimes u_2$ from $E_1 \otimes E_2$ into $F_1 \otimes F_2$ is continuous for the topologies of bi-equicontinuous convergence and can therefore be extended to a continuous linear mapping $u_1 \hat{\otimes} u_2$ from $E_1 \hat{\otimes} E_2$ into $F_1 \hat{\otimes} F_2$. If the u_i are topological isomorphisms, so is $u_1 \hat{\otimes} u_2$. If the E_i, F_i are Banach spaces, and if the u_i are metric isomorphisms, so is $u_1 \hat{\otimes} u_2$.

EXERCISE 4   Let E be a Hausdorff complete LCTVS. The following conditions are equivalent:

a) $E' \otimes F$ is dense in $L_c(E, F)$;
b) For every LCTVS F, $F' \otimes E$ is dense in $L_c(F, E)$;

c) For every LCTVS F with a set $\mathfrak{S}$ of bounded subsets, $F' \otimes E$ is dense for $\mathfrak{S}$-convergence in the space $K_{\mathfrak{S}}(F, E)$ of continuous linear mappings from F into E, which transform the $A \in \mathfrak{S}$ into relatively compact subsets of E;

d) for every Hausdorff complete LCTVS F, $F \,\widehat{\otimes}\, E$ can be identified with the space of weakly continuous linear mappings from F' into E which transform the equicontinuous subsets of F' into relatively compact subsets of E. Show that if E has these properties, every direct factor of E has them also.

7 Linear mappings from an LCTVS into certain function spaces. Mappings into a space of continuous functions

Let M be locally compact; consider the space $C(M)$ of continuous scalar functions on M, with the topology of uniform convergence on compact sets, and the space $C_0(M)$ of scalar functions "zero at infinity" with topology of uniform convergence (see Chapter 1, Section 9). For every $t \in M$, let $\varepsilon(t)$ be the linear form $f \mapsto f(t)$ on $C(M)$ which is clearly continuous, and the mapping $t \mapsto \varepsilon(t)$ from M into the dual of $C(M)$ (or of $C_0(M)$) is clearly continuous for the weak topology of the dual; it is easy to verify that it is even a homeomorphism (it is even a homeomorphism of M for the topology $\sigma((C(M))', \mathscr{K}(M))$ where $\mathscr{K}(M)$ is the space of continuous functions with compact support). Then the topology of $C(M)$ is a $\mathfrak{T}$-topology, where $\mathfrak{S}$ is the set of subsets of $(C(M))'$ of type $\varepsilon(K)$, where K runs through the set of compact subsets of M; the topology of $C_0(M)$ is the topology of uniform convergence on the subset $\varepsilon(M)$ of $(C_0(M))'$.

Let E be an LCTVS, u a continuous linear mapping from E into $C(M)$. Then its transpose u' is a weakly continuous mapping from $(C(M))'$ into E' which transforms the equicontinuous subsets into equicontinuous subsets (Chapter 2, Section 16) therefore $u' \circ \varepsilon$ is a continuous mapping f from M into E'_s, which transforms the compact subsets of M into equicontinuous subsets of E', defined explicitly by

$$(1) \qquad\qquad \langle x, f(t) \rangle = u(x)(t)$$

and we then see that u is known when f is known. Conversely, if f is a continuous mapping from M into E'_s which transforms compact subsets into equicontinuous subsets, then for every $x \in E$ the preceding formula defines a *continuous* scalar function on M (by the continuity of f); from this we get a linear mapping u from E into $C(M)$. This mapping is continuous, since it means (same reference) that when x' runs through a set $\varepsilon(K)$ (K a compact subset of M), then $x' \circ u$ runs through an equicontinuous subset of E', then by definition

$$\varepsilon(t) \circ u = f(t).$$

A similar reasoning is valid when we replace $C(M)$ by $C_0(M)$. Summing up:

THEOREM 5 *Let M be locally compact space, E an LCTVS. The continuous linear mappings from E into $C(M)$ (resp. into $C_0(M)$) correspond bijectively to the continuous mapping from M into E'_s which transform the compact subsets into equicontinuous subsets of E' (resp. to the continuous mappings "zero at infinity" from M into E'_s which map M into an equicontinuous subset of E'); this correspondence is the one given by formula (1) above.*

COROLLARY 1 *If E is barrelled, then we have the canonical isomorphisms*

$$L(E, C(M)) = C(M, E'_s)$$
$$L(E, C_0(M)) = C_0(M, E'_s).$$

In fact, every weakly compact subset of E' is equicontinuous, so that the supplementary condition on the continuous mappings from M into E'_s given in Theorem 5 becomes unnecessary. When E is only a TVS, we can only state:

$$L(E, C(M)) \subset C(M, E'_s)$$
$$L(E, C_0(M)) \subset C_0(M, E'_s).$$

Furthermore we verify immediately:

COROLLARY 2 *Let $\mathfrak{S}$ be a set of bounded subsets of E. Then the $\mathfrak{S}$-topology in $L(E, C(M))$ (resp. $L(E, C_0(M))$), is identical to the topology induced by the topology of compact (resp. uniform convergence on the space of mappings from M into $E'_{\mathfrak{S}}$ (where $E'_{\mathfrak{S}}$ stands for E' with a $\mathfrak{S}$-topology).*
(Proof: consider the neighborhoods of 0).

COROLLARY 3 *Let M be locally compact, E a Hausdorff LCTVS where the closed convex hull of a compact set is a compact set (for example, a quasi-complete space, see Chapter 2, Section 5, Exercise 4). Then we have isomorphisms*

$$C(M, E) = L(E'_c, C(M)), \quad C_0(M, E) = L(E'_c, C_0(M)),$$

where E'_c stands for E' with the topology of compact convergence.

In fact, by Mackey's theorem, the dual of E'_c is E, the result follows from Theorem 5 keeping in mind the fact that a continuous mapping from M into E weak, transforming compact subsets into compact subsets is already continuous for the given topology of E (since on a compact subset of E, the induced topology is identical to the induced weak topology), and on the other hand, an $f \in C(M, E)$ (or an $f \in C_0(M, E)$) evidently satisfies the conditions of Theorem 5 (that is, it transforms compact subsets into compact subsets—or transforms M into a relatively compact subset of E). Using the approximation theorem (Chapter 1, Section 9, Theorem 4). Corollary 3 can be rephrased:

COROLLARY 4   *Let M be a locally compact space, E an LCTVS which is separated and complete. Then we have the isomorphisms:*

$$C(M, E) \cong C(M) \,\widetilde{\otimes}\, E \quad C_0(M, E) \cong C_0(M) \,\widetilde{\otimes}\, E.$$

Besides, if E is a Banach space the second isomorphism is even a metric isomorphism

Let M be a set, E an LCTVS formed by scalar functions on M, with a topology finer than the topology of pointwise convergence (i.e. a vector subspace of k^M, with a locally convex topology which is finer than the induced topology). Therefore, for every $t \in M$, the linear form $\varepsilon(t)$ on E defined by

$$\langle \phi, \varepsilon(t) \rangle = \phi(t)$$

is continuous. Let F be a separated complete LCTVS, let u be a weakly continuous linear mapping from F' into E, then $f = u' \circ \varepsilon$ is a mapping from M into F given explicitly by

$$\langle f(t), y' \rangle = \langle uy', \varepsilon(t) \rangle = u(y')(t).$$

Then for $y' \in F'$, the function

$$t \mapsto f_{y'}(t) = \langle f(t), y' \rangle$$

is identical to uy', therefore, if y' runs through an equicontinuous subset of F', $f_{y'}$ runs through a weakly relatively compact subset of E. Conversely, if f is a mapping from M into F such that for every $y' \in F'$ the function $f_{y'}$ runs through a weakly relatively compact subset of E if y' runs through an equicontinuous subset of F', we shall show that the mapping $u: y' \mapsto f_{y'}$ from F' into E is weakly continuous. In fact, the restriction of u to every equicontinuous subset A of F', being continuous for the weak topology of A and the topology of pointwise convergence on E, is even weakly continuous since on the weak closure of $u(A)$ which is weakly compact, the weak topology is identical to the coarse separated topology of pointwise convergence on M. Thus, for every $x' \in E'$, the restriction of $x' \circ u$ to every equicontinuous subset A of F' is weakly continuous, therefore when F is complete, $x' \circ u$ will even be weakly continuous (Chapter 2, Section 14, Theorem 10), then u is weakly continuous. Summing up:

PROPOSITION 14   *Let M be a set, E an LCTVS formed by scalar functions on M with a topology finer than the topology of pointwise convergence. Let F be a complete Hausdorff LCTVS. Then the weakly continuous linear mappings from F' into E can be identified with the mappings f from M into F such that for every $y' \in F'$, the function $f_{y'} : t \mapsto \langle f(t), y' \rangle$ belongs to E and runs through a weakly relatively compact subset of E when y' runs through an equicontinuous subset of F'.*

COROLLARY   *In the preceding statement, if E is reflexive metrisable and complete, then the last condition on f is always satisfied.*

This means, in fact, that the mapping $y' \mapsto f_{y'}$ from F' into E transforms every weakly closed equicontinuous disk Λ of F' into a *bounded* subset of E, that is, induces a *continuous* linear mapping from the Banach space F'_Λ. But a priori this linear mapping is continuous for the topology on E of pointwise convergence in M, a Hausdorff topology which is coarser than the given topology on E. It follows from the closed graph theorem that this mapping is continuous.

Proposition 14, together with its corollary, can be applied in numerous cases. Choosing, for example, E to be the space of infinitely differentiable functions on an open subset M of $\mathbf{R}^n$, we find that the weakly continuous linear mappings from F' into E can be identified with the mappings from M into F which are "scalarly infinitely differentiable" (see the next section for the study of such functions), etc.

EXERCISE 1   Let M be a locally compact space, M_0 a closed subspace, E an LCTVS.

a) Show that the natural linear mapping from $C(M, E)$ into $C(M_0, E)$ assigning to every continuous function on M, its restriction to M_0, is a homomorphism from the first space onto a dense subspace of the second space.

b) Then conclude that if M is countable at infinity and E metrisable and complete we even obtain a homomorphism from the first space *onto* the second.

c) This is true for *every* complete locally convex space E if and only if the subspace $J(M_0)$ of $C(M)$ formed by functions zero on M_0 admits a topological supplement, or equivalently that the natural mapping $t \mapsto \varepsilon(t)$ from M_0 into the weak dual of $C(M_0)$ can be extended into a continuous mapping from M into the same space (for the sufficiency of the first condition use Corollary 3 of Theorem 5; for the converse, choose E to be the dual of $C_0(M_0)$ with the topology of compact convergence (for which the space is a complete LCTVS); then consider the continuous mapping $t \mapsto \varepsilon(t)$ from M_0 into E). This is the case if in particular M is metrisable and separable (Chapter 1, Section 11, Exercise 5).

d) Show that if the natural mapping from $C(M, E)$ into $C(M_0, E)$ is not onto, then the quotient of $C(M, E)$ by the closed subspace of functions zero on M_0 is not complete (although $C(M, E)$ is complete when E is complete).

EXERCISE 2   Let I be a set, consider the Banach space $l^\infty(I)$ of bounded families of scalars on I. If $J \subset I$, we identify $l^\infty(J)$ with the space $l^\infty(I)$ formed by elements whose coordinates $i \in \complement J$ are zero. If μ is a continuous linear form on $l^\infty(I)$, we denote its restriction to $l^\infty(J)$ by μ_J.

a) Show that if $J_1, \ldots, J_n$ are disjoint subsets of I, then

$$\| \mu_{J_1} \| + \cdots + \| \mu_{J_n} \| \leqslant \| \mu \|$$

b) For every continuous linear form μ on $l^\infty(I)$, let $\tilde{u}$ be its restriction to $c_0(I)$, identified with an element of $l^1(I)$ (Chapter 1, Section 9, Exercise 7); we can then consider it as a continuous linear form on $l^\infty(I)$ by the natural duality between $l^1(I)$ and $l^\infty(I)$, in general distinct from μ. Deduce from a) that if I is infinite, for every sequence (μ_n) of continuous linear forms on $l^\infty(I)$, there exists an infinite subset J of I such that $(\tilde{\mu}_n)_J = (\mu_n)_J$ for every n. (Construct by induction a decreasing sequence (J_n) of infinite subsets of I, such that

$$\| (\mu_n)_{J_n} \| \leqslant 1/n$$

for $k \leqslant n$, and consider an infinite subset J of I such that $J \cap \complement J_n$ is finite for every n).

c) From this conclude that if (μ_n) converges to zero in the weak dual of $l^\infty(I)$, then $(\tilde{\mu}_n)$ converges strongly to zero in $l^1(I)$ (proceed by contradiction, showing that if not we could find an $\varepsilon > 0$ and a sequence (A_n) of finite subsets of I, pairwise disjoint, such that

$$\sum_{i \in A_n} | \mu_n(i) | \geqslant \varepsilon$$

(if required replace (μ_n) by a subsequence, then consider the isomorphism of $l^\infty(N) - N$ is the set of integers which are positive or zero—into $l^\infty(I)$ given by

$$(\xi_n) \mapsto \sum_n \xi_n \phi_{A_n},$$

where ϕ_{A_n} is the characteristic function of A_n and where the series converges for the topology of pointwise convergence in $l^\infty(I)$; reduce to the case where $I = N$ and $A_n = \{n\}$. Using b) we can even suppose $\mu_n = \tilde{\mu}_n$ for every n, therefore that $\mu_n \in l^1$ for every n; then (μ_n) would be a weakly convergent sequence and not strongly convergent of the Banach space l^1, which is impossible (Chapter 2, Section 17, Exercise 4)).

d) Conclude from c) that if e_i is the ith coordinate form on c_0, then the sequence of e_i converges to 0 in the weak dual of c_0 but cannot be lifted to a sequence of continuous linear forms on l^∞ converging to zero in the weak dual of l^∞. A fortiori, c_0 has no topological supplement in l^∞.

e) Let M be the stone compactification of the space N of integers $\geqslant 0$, let M_0 be the compact subspace complement of N. Show that M, M^0 do not satisfy the conditions of Exercise 1c.

EXERCISE 3
a) Let M be a compact space, E a Banach space. Show that the space of compact linear mappings from E into $C(M)$ with the norm induced by $L(E, C(M))$, is canonically isomorphic to the normed space $C(M, E')$.

b) Then conclude that if E is a Banach space, F a normed vector subspace then every compact linear mapping u from F into $C(M)$ can be extended to a compact linear mapping from E into $C(M)$ of norm $\leqslant \| u \| + \varepsilon$ where $\varepsilon > 0$ is arbitrarily chosen (use Chapter 1, Section 14, Exercise 2)

c) Generalize to the case where E is any LCTVS.

EXERCISE 4 Show that we can find a closed subspace F of $E = l^1$ and a sequence in E/F converging weakly to zero which does not come from a sequence in E converging weakly to zero (recall that every weakly convergent sequence of l^1 is strongly convergent—Chapter 2, Section 17, Exercise 4); show also that there exist separable Banach spaces, such as c_0, which admit weakly convergent sequences which are not strongly convergent; show finally that every separable Banach space is isomorphic to a quotient space of l^1—(Chapter 1, Section 14, Exercise 1).

8 Differentiable vectorial functions

Let U be open in R^n, E a Hausdorff LCTVS (in this section we consider *Hausdorff* LCTVS only), f a mapping from U into E. The notion of differentiability of f in a point $t \in U$, with respect to one of the variables (or more generally, in the direction of a given vector) can be defined as in the scalar case, and from this follows the notion of a mapping which is m times continuously differentiable, from U into E. We now develop some useful lemmas.

LEMMA 1 *A mapping f from $U \subset R^n$ into the weak completion E'^* of E is differentiable at $t \in U$ if and only if for every $x' \in E'$, the function $f_{x'} = \langle f(t), x' \rangle$ is differentiable at t; then we have for every $x' \in E'$*

$$(1) \qquad\qquad \langle f'(t), x' \rangle = f_{x'}(t).$$

In fact, we must state that

$$\frac{1}{h}\left(f(t + h) - f(t)\right)$$

considered as a linear form on E' converges simply to a limit if $h \to 0$ (this limit will be automatically linear, therefore it will be an element of E'^*, the derivative of f in t). This means that for every $x' \in E'$ the scalar product

$$\left\langle \frac{1}{h}\left(f(t + h) - f(t)\right), x' \right\rangle = \frac{1}{h}\left(f_{x'}(t + h) - f_{x'}(t)\right)$$

tends to a limit (that will be $\langle f'(t), x' \rangle$, i.e. that for every $x' \in E'$ the function $f_{x'}$ is differentiable in t whence (1)).

COROLLARY *A mapping f from the open set $U \subset R^n$ into the weak completion E'^* of E is m times continuously differentiable if and only if f is scalarly m times continuously differentiable (by this we mean that for every $x' \in E'$, the function $f_{x'}$ has the property considered)*

LEMMA 2 *The mapping f from U into E is m times continuously differentiable in the weak completion E'^* of E if and only if it is so in E'' with the weak dual topology.*

The sufficiency is trivial; for the converse we start with $m = 1$ and $U \subset R$ and we note that if the limit of $(1/h)(f(t + h) - f(t))$ exists in E'^* weak, this limit is also the limit of a *sequence* in E (corresponding to $h = 1/n$ with n an integer > 0), which will be a weak Cauchy sequence, therefore bounded. It follows (Mackey's theorem) that the limit belongs to E''.

LEMMA 3 *Let f be a mapping from U into E which is continuously differentiable and whose first partial derivatives are continuous for a locally convex topology T on E with a closed fundamental system of neighborhoods of 0. Then f is also continuously differentiable for T.*

The hypothesis on T means that it is a $\mathfrak{S}$-topology, where $\mathfrak{S}$ is a set of weakly closed subsets of E'. We need only consider the case of one variable and it suffices to show that $(1/h)(f(t+h) - f(t))$, tending to $f'(t)$ for the initial topology T_0 of E, tends also to the same limit for the topology T, that is, that for every $A \in \mathfrak{S}$, its scalar product with the $x' \in A$ tends to $\langle f'(t), x' \rangle$ *uniformly* when x' runs through A. Now this scaler product can also be written

$$\frac{1}{h}\left(f_{x'}(t+h) - f_{x'}(t)\right) = \frac{1}{h}\int_t^{t+h} f'_{x'}(s)\, ds,$$

its difference with $\langle f'(t), x' \rangle = f_{x'}(t)$ can be written

$$\frac{1}{h}\int_t^{t+h} \left(f'_{x'}(s) - f'_{x'}(t)\right) ds,$$

and it tends, in fact, to zero uniformly when x' runs through an $A \in \mathfrak{S}$, since $f'_{x'}(s)$ tends to $f'_{x'}(t)$ for $s \to t$, uniformly when x' runs through an $A \in \mathfrak{S}$ (this is a statement of the continuity of $f'(s)$ for the topology T).

COROLLARY *T being as above, if f is m times continuously differentiable and its derivatives of order m are continuous for T, then f is also m times continuously differentiable for T.*

Proof This is immediate starting from Lemma 3 by induction on m.

LEMMA 4 *Let F be an LCTVS, E a quasicomplete vector subspace, f a mapping from the open set $U \subset R^n$ in E which is m times continuously differentiable, into F. Then f is m times continuously differentiable in E.*

The proof is analogous to Lemma 2 since the values in F of the derivatives of f will be in the closure of the *bounded* subsets of E.

LEMMA 5 *Let f be a mapping scalarly continuously differentiable from U into E. Then f is continuous even for the topology of uniform convergence on the strongly bounded subsets of E'.*

The proof is easier if we use the notion of a weak integral, the elementary properties only being needed here. We have

$$f(t+h) - f(t) = \int_t^{t+h} f'(s)\, ds$$

(we integrate the continuous function f' with values in E'' weak), then we obtain

$$f(t+h) - f(t) \in hK \quad \text{for } h \text{ sufficiently small,}$$

where K is the weakly closed convex hull of the set K_0 which is the image of a fixed compact interval of center t, contained in U. The set K_0 is weakly compact therefore K is weakly bounded, therefore

$A = K \cap E$ is a *bounded* subset of E and we have

$$f(t + h) - f(t) \in h \cdot A.$$

From this it follows that $(f(t + h) - f(t))$ tends to zero for the topology of uniform convergence on the strongly bounded subsets of E'. (We have supposed, for the sake of simplicity of notation, $U \subset \mathbf{R}$, a fact which is not essential.) Notice that Lemma 3 could also be proved using this integral method.

PROPOSITION 15   *Let U be open in $\mathbf{R}^n$, E a quasi-complete LCTVS, f a mapping from U into E. The mapping f is m times continuously differentiable if and only if it is scalarly m times continuously differentiable and its mth partial derivatives (that exist in E''—see Lemmas 1 and 2) are continuous for the natural topology of E'' (the topology of uniform convergence on the equicontinuous subsets of E').*

Explicitly, these conditions mean that the $f_{x'}$ are m times continuously differentiable, and that the mth derivatives run through an *equicontinuous* set of functions on U, when x' runs through an equicontinuous subset of E' (this condition permits the consideration of only the scalar components $f_{x'}$).

Proof   Under the indicated conditions f is m times continuously differentiable in E'' weak (Lemmas 1 and 2) and so is also for the natural topology of E'' (Lemma 3), therefore also in E itself (Lemma 4).

COROLLARY 1   *E being quasi-complete, if the mapping f from U into E is scalarly m times continuously differentiable, it is $m - 1$ times continuously differentiable.*

In fact, by Lemma 5 the derivatives of order $m - 1$ of f into E'' weak are continuous for the topology of uniform convergence on the subsets of E' bounded for the strong topology of E' associated with the dual system (E', E''), then a fortiori for the topology of uniform convergence on the weakly compact equicontinuous disks of E' (therefore complete and bounded for the strong topology on E' associated with (E', E'')) i.e. the natural topology of E''. The corollary then follows from Proposition 15. We conclude from this:

COROLLARY 2   *Let E be a quasi-complete LCTVS, f a mapping from an open set U of $\mathbf{R}^n$ into E. The mapping f is indefinitely continuously differentiable if and only if f is scalarly indefinitely continuously differentiable.*

PROPOSITION 16   *Let E be a complete LCTVS, let U be open in $\mathbf{R}^n$. Then* *the space $\mathscr{E}^{(m)}(U, E)$ of m-times continuously differentiable functions from U into E with its natural topology (compact convergence of f and its derivatives of order $\leqslant m$) can be identified with the space of weakly continuous linear mappings from E' into $\mathscr{E}^{(m)}(U)$ which transform the equicontinuous subsets into relatively compact subsets of $\mathscr{E}^{(m)}(U)$, this space of mappings being endowed with the topology of uniform convergence on the equicontinuous subsets of E'.*

Let $f \in \mathscr{E}^{(m)}(\cup, E)$. Then f defines a linear mapping $x' \mapsto f_{x'}$ from E' into $\mathscr{E}^{(m)}(U)$, which is weakly continuous by Section 7, Proposition 14. We now show that it transforms an equicontinuous subset A of E' into a relatively compact subset of $\mathscr{E}^{(m)}(U)$. From Chapter 1, Section 10, Proposition 18, it suffices to show that the set of derivatives of order m of the $f_{x'}(x' \in A)$ is an equicontinuous set of functions on U_n this means that the mth derivatives of f in E are continuous mappings. Conversely, given a weakly continuous mapping u from E' into $\mathscr{E}^{(m)}(U)$, which transforms the equicontinuous subsets into relatively compact subsets, we deduce a mapping f from U into E such that $\langle f(t), x' \rangle = ux'(t)$ for $t \in U$, $x' \in E'$. This f then satisfies the conditions of Proposition 15 and belongs to $\in^{(m)}(\cup, E)$. Finally, we verify trivially that the topology of $\in^m(\cup, E)$ corresponds to the topology of uniform convergence on the equicontinuous subsets of E. In this reasoning we could have supposed that $m = +\infty$, and dropped the condition of compactness using Corollary 2 of Proposition 15. We thus obtain

COROLLARY   *E being a complete LCTVS, $\mathscr{E}(U, E)$ can be identified with the space of weakly continuous linear mappings from E' into $\mathscr{E}(U)$, with the topology of uniform convergence on the equicontinuous subsets of E'.*

This is also the corollary of Proposition 14 taking into consideration Proposition 15, Corollary 2.

K

CHAPTER 4

Study of some special classes of spaces

PART 1   INDUCTIVE LIMITS, ($\mathscr{LF}$) SPACES

1 Generalities

DEFINITION 1   *Let E be a vector space, (E_i) a family of LCTVS, and for every i let u_i be a linear mapping from E_i into E. We call the finest locally convex topology on E for which the u_i are continuous the inductive limit topology of the E_i (by the u_i). With this topology E is called the inductive limit of the E_i (by the u_i).*

The existence of such a topology is immediate as a locally convex topology T on E permits the u_i to be continuous if and only if for every disked neighborhood V of 0 for T, the $u_i^{-1}(V)$ are neighborhoods of 0 in the E_i; clearly the set of all absorbing disks in E having this property is a system of disked neighborhoods of 0 for a locally convex topology on E which is the finest among the topologies considered. We have shown the first part of

PROPOSITION 1   *Let E be the inductive limit of the E_i by the u_i.*

1) *An absorbing disk V in E is a neighborhood of 0 if and only if for every i, $u_i^{-1}(V)$ is a neighborhood of 0 in E_i.*

2) *If the $u_i(E_i)$ generate E, we obtain a fundamental system of neighborhoods of 0 by taking the convex hulls*

$$\Gamma\left(\bigcup u_i(V_i)\right),$$

where, for every i, V_i runs through a given fundamental system of neighborhoods of 0 in E_i.

The second part is an immediate consequence of the first. Notice that in 1), if the $u_i(E_i)$ generate E, it is unnecessary to suppose a priori V absorbing, since it will be so automatically.

COROLLARY 1   *An inductive limit of bornological (resp. barrelled, resp. quasi-barrelled) spaces is bornological (resp. barrelled, resp. quasi-barrelled).*

In fact, if V is a disk in E which is bornivorous (resp. closed and absorbing, resp. closed and bornivorous), it is also balanced and the $u_i^{-1}(V)$ are clearly also bornivorous (resp. . . . .) therefore they are neighborhoods of 0 by the hypothesis on the E_i, and V is itself a neighborhood of 0, hence the conclusion.

Very often, the family of indices I (implicit above) is ordered, increasingly directed and for $i \leqslant j$ we define a continuous linear mapping u_{ij} from E_i into E_j such that $i \leqslant j \leqslant k$ implies $u_{ik} = u_{jk}u_{ij}$ and $i \leqslant j$ implies $u_i = u_j u_{ij}$.

We verify immediately that the inductive limit topology on E does not change if we replace I by a set $I' \subset I$ cofinal to I. We can reduce the general case to the one just considered. In fact, for the general case, for every finite subset $J \subset I$, let

$$E_J = \prod_{i \in J} E_i,$$

and let u_J be the linear mapping from E_J into E coinciding with the u_i on the E_i; since this mapping is clearly continuous when E has the inductive limit topology of the E_i, it follows that the latter is also identical to the inductive limit of the E_J by the u_J; clearly, the family (E_J) forms a transitive system of the type considered above. Notice furthermore that the inductive limit topology of E does not change when we replace the E_i by their quotients by the kernel of the u_i, which leads us to the case where the u_i are injective and consequently the E_i are identified with vector subspaces of E having their own locally convex topologies T_i. When the E_i form a transitive system we see that after passage to the quotient we have subspaces E_i of E such that $i \leqslant j$ implies $E_i \subset E_j$ and such that the identity mapping from E_i into E_j is continuous (for the given topologies of E_i, E_j). Very often, the union of the $u_i(E_i)$ generates E (therefore, if the E_i form a transitive family, the union of the $u_i(E_i)$ is identical to E).

Notice the transitivity property resulting from the definition: If each E_i is itself an inductive limit of $(E_{i,\alpha})_{\alpha \in A}$ then E is also the inductive limit of $(E_{i,\alpha})_{\alpha \in A}$, $i \in I$, by the $u_i \circ u_{i,\alpha}$.

PROPOSITION 2   *Let E be the inductive limit of (E_i) of LCTVS by (u_i), E generated by the $u_i(E_i)$. Let F be an arbitrary LCTVS.*

1) *A linear mapping v from E into F is continuous if and only if for every i $v \circ u_i$ is continuous. More generally, let M be a set of linear mappings from E into F; M is equicontinuous if and only if for every i, $M \circ u_i$ (formed by the $v \circ u_i$ with $v \in M$) is an equicontinuous set of mappings from E_i into F.*

2) *Let, for every i, $\mathfrak{S}_i$ be a set of bounded subsets of E_i; let $\mathfrak{S}$ be the union of bounded subsets $u_i(\mathfrak{S}_i)$ of E. Then the $\mathfrak{S}$-topology in the space of linear*

mappings from E into F is the coarsest topology for which the mappings $v \mapsto v \circ u_i$ are continuous when the space of linear mappings from E_i into F has the $\mathfrak{S}_i$-topology. In particular, a set M of linear mappings from E into F is bounded for the $\mathfrak{S}$-topology if and only if for every i the set $M \circ u_i$ of linear mappings from E_i into F is bounded for the $\mathfrak{S}_i$-topology.

Proof 2) is trivial (F can be any topological space, the same for the mappings; the vector structures and the topology of E are not considered). For 1) we notice that if M is equicontinuous, for every neighborhood V of 0 in F, $M^{-1}(V)$ is a neighborhood of 0 in E; since it is already a disk, this means (Proposition 1) that for every i, $u_i^{-1}(M^{-1}(V))$ is a neighborhood of 0 in E_i. Now this set is also $M_i^{-1}(V)$ where $M_i = M \circ u_i$; we must express that for a given i, $M_i^{-1}(V)$ is a neighborhood of 0 in E_i for any disked neighborhood V of 0 in F, i.e. that M_i is equicontinuous.

Some questions arise concerning a space which is an inductive limit, which often receive negative answers, even for the inductive limits of a sequence of Banach spaces, and which often present serious difficulties.

1) Is E complete when the E_i are complete?

2) Does every bounded (resp. compact, resp. weakly compact) subset of E come from a bounded subset (resp. . . . .) of a space E_i? (In this question, we can suppose that the (E_i) already form a transitive system.) In particular, if we suppose the E_i reflexive, (or Montel spaces) is E reflexive (or a Montel space). (For an affirmative answer it would suffice to know that E is Hausdorff and that every bounded subset of E comes from a bounded subset of an E_i.) Notice that even if the E_i are Hausdorff it is possible that E is not Hausdorff (see Section 4, Exercise 2) but in practice, we make sure that E is Hausdorff since it suffices for this purpose to find on E a Hausdorff locally convex topology for which the u_i are continuous. We remark that in practice the difficulties which we encounter in inductive limits are the "converse" of those met in projective limits (the coarsest topology for which . . .); here it is nearly always easy to show that the space is complete, and to determine whether its bounded subsets are weakly compact or compact (the reader will recall the corresponding statements), and in particular to recognize it as either a reflexive or a Montel space. But it is difficult to recognize if the space is bornological or barrelled, and we have no good criterion as in Proposition 2 above for the equicontinuous sets of linear forms, etc.

2 Examples

In all of these examples, the E_i naturally form a transitive system and the union of the $u_i(E_i)$ is E. We verify trivially in examples c), d), e), f), that the E obtained in each case is Hausdorff, using the general remark stated in Section 1.

a) Let E be an LCTVS, F a vector subspace. Then E/F is the inductive limit of E by the canonical mapping of E onto E/F.

b) The locally convex space E is bornological if and only if it is an inductive limit of normed spaces. In fact, it is sufficient by Proposition 1, Corollary (as a normed space is bornological by Chapter 3, Section 4, Theorem 3) and it is necessary as we verify immediately that E is bornological if and only if its topology is the inductive limit of the spaces E_A (where A runs through the closed and bounded disks of E) by the identity mappings of these spaces into E. (Then if E is Hausdorff and quasi-complete, the E_A are complete, therefore E is bornological if and only if it is an inductive limit of a family of *Banach spaces*.)

c) Let E and F be ($\mathscr{F}$) spaces, let $\Gamma(E, F)$ be the space of compact linear mappings from E into F. For every disked neighborhood V of 0, let E_V be the normed space associated to the gauge, semi-norm of V; there exists a natural linear bijection from $\Gamma(E_V, F)$ into $\Gamma(E, F)$ which is the union of the images of these spaces. Equip $\Gamma(E_V, F)$ with a topology of bounded convergence for which it is a *closed* subspace of the space $L_b(E_V, F)$ (Chapter 3, Section 1, Proposition 2; the *compact* linear mappings from a semi-normed space into a space F are those transforming the *unit ball* into a relatively compact subset of F). Since $L_b(E_V, F)$ is clearly an ($\mathscr{F}$) space, so is $\Gamma(E_V, F)$ (we thus have the right topology on this space). We can then equip $\Gamma(E, F)$ with the inductive limit topology of the spaces $\Gamma(E_V, F)$ of type ($\mathscr{F}$). This limit topology does not change if V runs through a fundamental system of disked neighborhoods of 0, for example a fundamental *sequence* (V_n) of neighborhoods. Thus, the space $\mathscr{C}(E, F)$ of compact linear mappings of a space ($\mathscr{F}$) into another appears as an inductive limit of a sequence of spaces of type ($\mathscr{F}$). There is an analogous construction for the space of bounded linear mappings from E into F.

d) Let M be a locally compact space, let $\mathscr{K}(M)$ be the space of continuous numerical functions with compact support on M. For every compact set $K \subset M$ we denote by $\mathscr{K}_K(M)$ the space of functions with support $\subset K$ with the uniform norm topology (i.e. the topology of uniform convergence). Thus it becomes a Banach space. $\mathscr{K}(M)$ is the union of these subspaces and we can equip it with an inductive limit topology. If M is countable at infinity, we can choose some increasing sequence (K_n) of compact set, then $\mathscr{K}(M)$ is the inductive limit of a *sequence* of Banach spaces.

e) We proceed likewise for the space $\mathscr{D}(U)$ of indefinitely differentiable functions, with compact support, on an open set $U \subset \mathbf{R}^n$; this space is the inductive limit of a sequence of subspaces of type $(\mathscr{F})$.

f) Let, first of all, U be open in $\mathbf{R}^n$ and consider the space $H(U)$ of holomorphic functions on U which is a *closed* vector subspace of the space $C(U)$ of continuous complex functions on U (Weierstrass theorem), and has the induced topology for which it is a space $(\mathscr{F})$. Let A be a subset of $\mathbf{C}^n$ and $H(A)$ the set of equivalence classes of holomorphic functions defined in an open neighborhood of A; two functions are equivalent if they coincide on a neighborhood of A. The set $H(A)$ has an obvious linear structure, and if $B \supset A$ we have a natural linear mapping $\phi_{B,A}$ from $H(B)$ into $H(A)$ (the *restriction*). If A is open, we find the space already defined. We can then equip $H(A)$ with the inductive limit topology of the $H(U)$ corresponding to the open neighborhoods U of A, by the linear mappings $\phi_{U,A}$. It suffices to make U run through a fundamental system of neighborhoods of A. In particular, if A is compact, we can choose a fundamental sequence (U_n) of open neighborhoods of A, therefore $H(A)$ appears then as an inductive limit of a sequence of $(\mathscr{F})$ spaces.

3 Strict inductive limits

DEFINITION 2 *E is a strict inductive limit of a family (E_i) of LCTVS if the E_i form an increasingly directed set of vector subspaces of E whose union is E, each E_i with a Hausdorff locally convex topology and $E_i \subset E_j$ implying that E_i is a closed topological vector subspace of E_j (in particular, its topology is induced by E_j). This is the case in Examples* d), e), *of Section* 2.

We still do not know if in this case the questions asked at the end of section 1 have an affirmative answer (see Exercise 7 below). We shall see that it is so in the most important case where the index set I is countable. In this case, the remarks made in Section 1 show that we can suppose (E_i) to be an *increasing sequence* of vector subspaces of E, whose union is E, with given Hausdorff topologies such that the topology of E_i is induced by that of E_{i+1}, and such that E_i is closed in E_{i+1}.

PROPOSITION 3 *Let E be a strict inductive limit space of an increasing sequence of subspaces E_i. Then the E_i are closed in E, E induces on each E_i the* *given topology of E_i (in particular E is Hausdorff), the bounded subsets of E are those contained and bounded in some space E_i. The space E is complete if and only if the E_i are complete.*

The proof rests on the

LEMMA   *Let F be an LCTVS, G a closed vector subspace, V an open disk in G, $x \in \complement G$. Then, there exists an open disk W in F such that $x \in \complement W$, $W \cap G = V$.*

Let W_0 be an open disk in F sufficiently small so that $x + W_0$ does not meet G and such that $W_0 \cap F \subset V$. The reader is invited to verify that $W = \Gamma(W_0 \cap V)$ satisfies the condition.

In order to show that E induces on E_i the given topology of E_i we must show that for every open disk V_i in E_i there exists a disked neighborhood of 0 in E such that $V \cap E_i = V_i$. But by the lemma, we construct by induction a sequence of disks $V_{i+1}, \ldots, V_j, \ldots$ where V_j is an open disk in E_j inducing V_{j+1} on E_{j-1}. Then $V = \bigcup V_j$ is the required neighborhood. We now show that E_i is closed in E that is, for $x \in \complement E_i$ there exists a continuous linear form on E, zero on E_i but not zero on x. In fact we have $x \in E_j$ for some j and since E_i is closed in E_j, there exists a continuous linear form on E_j zero on E_i but not on x. Since E_j is a topological vector subspace of E, this form can be extended into a continuous linear form defined on all of E which will satisfy the condition. For the assertion relative to the bounded sets we must show that if A is a bounded subset of E then $A \subset E_i$ for some i. If this were not the case we could construct by induction a strictly increasing sequence of indices (i_n) and a sequence (x_n) in E such that

$$x_n \in A \cap E_{i_n} \cap \complement E_{i_{n-1}}.$$

Then by the lemma we have a sequence of open disks V_n in the E_{i_n} such that $V_{n-1} = V_n \cap E_{i_{n-1}}$, $x_n \notin n V_n$. The union V of V_n is then a neighborhood of 0 in E, inducing the V_n on the E_{i_n} such that $x_n \notin n V_n$ for every n. It follows that the sequence (x_n), therefore A, cannot be bounded.

If E is complete so are its closed subspaces E_i. Conversely, supposing that the E_i complete, we shall show that E is complete, i.e. (Chapter 2, Section 14, Theorem 10) that every linear form x on E' whose restrictions to the equicontinuous subsets are weakly continuous, is already weakly continuous. First, x is zero on an orthogonal set E_i^0 for some i, as otherwise for every i there exists an $x_i' \in E_i^0$ such that $\langle x, x_i' \rangle = 1$, and the sequence of x_i' would be equicontinuous (Proposition 1) and would converge weakly to 0 without $\langle x, x_i' \rangle$ tending to 0, which is absurd. Thus x is zero on a space E_i^0, therefore it comes from a linear form y on the quotient E'/E_i^0, which quotient can be identified with the dual E_i' of E_i. This linear form is restricted to the equicontinuous subsets of E_i' which are weakly continuous; in fact, an equicontinuous subset of E' is contained in the image $\phi(A)$ of an equicontinuous subset of E' (Chapter 2, Section 15, Proposition 20) which we can suppose weakly compact, where ϕ is the canonical mapping from E' onto E_i'. Now the restriction of y to $\phi(A)$ weak is continuous if and only if $y \circ \phi = x$ is continuous on A weak which is in fact the case. Since E_i is complete we conclude that y is weakly continuous and so is $x = y \circ \phi$.

4. Direct sums

DEFINITION 3   *Let (E_i) be a family of LCTVS E_i. The topological direct sum of the spaces E_i, is the algebraic direct sum $E = \underset{i}{\Sigma E_i}$ with the topology which is the inductive limit of the subspaces E_i (with their given topology).*

When the index set is finite, we have the product topology.

The construction of general inductive limits can be reduced to the construction of topological direct sums and quotients, by

PROPOSITION 4   *Let E be an LCTVS, the inductive limit of a family (E_i) of LCTVS by linear mappings u_i such that E is generated by the union of the $u_i(E_i)$. Then E is isomorphic to a quotient space of the topological direct sum of the E_i.*

Let F be this last space, let u be the linear mapping of F into E which coincides with u_i on E_i; it is a linear mapping from F *onto* E, therefore it defines a bijective linear mapping from a quotient of F onto E. The quotient topology on E is the inductive limit topology of F by u (Section 2, Example a)), therefore (by transitivity of inductive limits, see Section 1) it is identical to the inductive limit topology of the E_i by the u_i induced by u on the E_i; hence the conclusion.

PROPOSITION 5   *Let E be the topological direct sum of a family $(E_i)_{i \in I}$ of Hausdorff LCTVS. Then the E_i are closed in E and E induces in them the given topology; more generally, for every subset J of I, the space E_J, the topological direct sum of the E_i with $i \in J$, is a closed topological vector subspace of E. The bounded subsets of E are the subsets which are contained and bounded in the direct sum of a finite number of factors E_i. The space E is complete if and only if the E_i are complete.*

A priori, the direct sum topology of E_J is finer than the topology induced by E, but the natural projection of E onto E_J is a continuous linear mapping from E into E_J (as its restrictions to E_i, $i \in I$, are continuous), thus E_J is a closed topological vector subspace of E. Let A be a bounded subset of E; if A were not contained in any E_J with J finite, we could construct by induction an increasing sequence of finite subsets $J_1, J_2, \ldots, J_n, \ldots$ of I and a sequence of points $x_1, \ldots, x_n, \ldots$ of A, such that

$$x_n \in E_{J_{n+1}} \cap \complement E_{J_n}.$$

We let J be the union of the J_n, and verify that E_J is the strict inductive limit of the E_{J_n}. But then $A \cap E_J$ would be a bounded subset of E_J which is not contained in any of the E_{J_n}, a contradiction to Proposition 3. If E is complete, the E_i are complete as they are closed subspaces of E. For the converse, we need first

PROPOSITION 6   *The dual of the topological direct sum ΣE_i can be identified with the product $\Pi E'_i$ of duals of E_i, the equicontinuous subsets of the dual are those contained in a product ΠA_i, where for every i, A_i is an equicontinuous subset of E'_i.*

This is a trivial consequence of Proposition 2, 1.

COROLLARY   *Let for every i, $\mathfrak{S}_i$ be a set of bounded subsets of E_i, let $\mathfrak{S}$ be the set of bounded subsets of E, the union of the $\mathfrak{S}_i$. Then on $E' = \Pi E'_i$ the $\mathfrak{S}$-topology is identical to the product topology, when the E'_i have a $\mathfrak{S}_i$-topology. In particular, the weak (resp. strong) topology on E' is the product of the weak (resp. strong) topologies of the E'_i.*

The general assertion is a particular case of Proposition 2, 2. The particular cases follow as we can replace $\mathfrak{S}$ by the set of convex hulls of finite unions of elements of $\mathfrak{S}$ and then apply the characterization of bounded sets of E given in Proposition 5.

We can now show that if the E_i are complete, E is complete, i.e. that every linear form x on $E' = \Pi E'_i$, whose restrictions to the equicontinuous subsets are weakly continuous, is weakly continuous (Chapter 2, Section 14, Theorem 10). Let x_i be the restriction of x to E'_i, then its restrictions to the equicontinuous subsets of E'_i (considered as the dual of E_i) are clearly weakly continuous, therefore, E_i being complete, we have $x_i \in E_i$. We now show that all the x_i except for a finite number are zero: if not, there would exist an infinite sequence of distinct indices $i_1, \ldots, i_n, \ldots$ and of elements $x'_{i_n} \in E'_{i_n}$ such that

$$\langle x, x'_{i_n} \rangle = \langle x_{i_n}, x'_{i_n} \rangle = 1,$$

whence it is immediate that (x'_{i_n}) is an equicontinuous sequence tending weakly to 0 in E', which is absurd, since $\langle x, x'_{i_n} \rangle$ would have to tend to zero. Let X be the element of E whose components are the x_i; it remains to be shown that $x = X$. These two forms coincide on the subspace E'_i of $\Pi\, E'_i$ and are restricted to the equicontinuous subsets of $\Pi\, E'_i$ which are weakly continuous, thus it suffices to show that every

$$x' = (x'_i) \in \Pi\, E'_i$$

is in the weak closure of an equicontinuous subset of $\Pi\, E'_i$ contained in $\Sigma\, E'_i$. The set of projections x'_J of x' to the finite direct sums E'_J corresponding to the finite subsets J of I is then clearly equicontinuous, and x' is a weak limit of the x'_J with respect to the filter of increasing sections in the set of finite subsets of I, hence the conclusion.

The following is the dual of Proposition 6:

PROPOSITION 7   Let (E_i) be a family of LCTVS, let for every i, $\mathfrak{S}_i$ be a set of bounded disks of E_i; let $\mathfrak{S}$ be the set of subsets of $\Pi\, E_i$ of type $\Pi\, A_i$ with $A_i \in \mathfrak{S}_i$ for every i. Then on the dual $\Sigma\, E'_i$ of the topological vector product $\Pi\, E_i$ (see Chapter 2, Section 15, Proposition 22) the $\mathfrak{S}$-topology is identical to the direct sum topology of the $\mathfrak{S}_i$-topologies on the E_i.

This follows immediately from the following formula, valid for every family (A_i) of disks in the E'_i:

$$(\Pi\, A_i)^0 = \bar{\Gamma}(U(A_i)^0)$$

the verification of which is left to the reader. By the characterization of the bounded subsets of $\Pi\, E_i$, we find the

COROLLARY   The strong dual of $\Pi\, E_i$ can be identified with the topological direct sum of the strong duals of the E_i.

EXERCISE 1   Generalize Proposition 6 and its corollary to the space of continuous linear mappings from a topological direct sum into an LCTVS. Generalize also Proposition 7 and Chapter 2, Proposition 22 to the space of continuous linear mappings from a topological vector product of LCTVS into an LCTVS F which is *semi normed* (in particular a Banach space). Show that the result obtained is false when F is any LCTVS (take $F = E$).

EXERCISE 2   Let H be an infinite dimensional separable Banach space. Let (H_n) be an increasing sequence of vector subspaces of finite dimension whose union N is dense in H, let $E_n = H/H_n$, $E = H/N$, u_n being the natural mapping from E_n onto E. Show that the inductive limit topology of the sequence of Banach spaces E_n by the u_n is the coarsest topology on E (i.e. every non-empty open set is identical to E). Let F be an algebraic supplement of N, let F_n be the space F whose norm is the inverse image of the norm of E_n by the natural mapping of F into E_n; show that on F, the inductive limit topology of the F_n (with respect to the identity mappings) is still the coarsest topology on F. (Show that every continuous linear form on E or F is zero.)

EXERCISE 3   Let E be a vector space, (E_i) an increasing sequence of vector subspaces of E with Hausdorff locally convex topologies such that the identity mapping from E_i into E_{i+1} is weakly compact (Chapter 2, Section 18, Definition 14) for every i. Show that the space E, with the inductive limit topology of the E_i is Hausdorff and that it can also be considered as the inductive limit of an analogous sequence of *Banach* spaces.

EXERCISE 4 (G. Köthe)   Let for every n, (a_{ij}^n) be the double sequence defined by: $a_{ji}^n = j^n$ for $i \leqslant n$, $a_{ij}^n = i^n$ for $i > n$. Let

$$c_0 = c_0(\mathbf{N} \times \mathbf{N}) \qquad l^\infty = l^\infty(\mathbf{N} \times \mathbf{N}),$$

where $\mathbf{N}$ is the set of natural numbers, let $a^n . c_0$ (resp. $a^n . l^\infty$) be the space of products $a^n x$ with $x \in c_0$ (resp. $x \in l^\infty$) with the norm topology deduce from the topology of c_0 (resp. l^∞) by transport of structure.

a) For every double sequence x and every subset J of $\mathbf{N} \times \mathbf{N}$ let x_J be the double sequence whose coordinates are identical to those of x on J and zero on $\complement J$. Let E resp. F) be the union of $a^n . c_0$ (resp. $a^n . l^\infty$) with the inductive limit topology. Show that for every $x \in F$, the sequence of x_J where J runs through the finite subsets of $\mathbf{N} \times \mathbf{N}$ is a Cauchy sequence in E and that it tends to the limit x in F.

b) From this conclude that E is not complete, and similarly for the sequences (notice that the double sequences whose coordinates are all equal to 1 is in F, but not in E).

EXERCISE 5   Let E be an LCTVS, a strict inductive limit of an increasing sequence of subspaces E_n.

a) If $\mathfrak{S}$ (resp. $\mathfrak{S}_n$) is the set of equicontinuous subsets of E' (resp. E_n') then for every $\mathfrak{S}$-absorbing disk U in E' (see Chapter 2, Section 3, Definition 1) there exists an n such that $U \supset (E_n)^0$, and then the set U_n of $x' \in (E_n)'$ whose inverse image in E' is contained in U is a $\mathfrak{S}$-absorbing disk.

b) Conclude from this that if the E_n are quasi-barrelled and if their strong duals are bornological, then E is quasi-barrelled and its strong dual is bornological.

c) Let E be a quasi-barrelled LCTVS. Show that if E' strong is quasi-barrelled, it is also barrelled (therefore under the conditions of b) E' strong is also barrelled).

EXERCISE 6 Deduce from Exercise 4 and Proposition 4 that there exists a closed vector subspace H of the direct sum L of a sequence of spaces isomorphic to c_0 such that L/H is not complete (even if L is complete by Proposition 5). Show that this remains true if we replace c_0 by l^1 (use the fact that c_0 is isomorphic to a quotient space of l^1— Chapter 1, Section 14, Exercise 1).

EXERCISE 7 Let $E = \Pi E_i$ be the topological vector product of a family of LCTVS E_i.

a) Show that the weakly (resp. strongly) bounded subsets of $E' = \Sigma E_i'$ are the subsets contained in the sum of a finite number of weakly (resp. strongly) bounded subsets of E_i'.

b) Conclude that E is barrelled (or quasi-barrelled) if and only if the E_i are.

EXERCISE 8 Let $E = \Sigma E_i$ be the topological direct sum of a family of LCTVS E_i. The space E is bornological (resp. barrelled, resp. quasi-barrelled, resp. reflexive, resp. of type $(\mathscr{M})$) if and only if the E_i are bornological (resp. . . .).

5 $(\mathscr{L}\mathscr{F})$ spaces

DEFINITION 4 An $(\mathscr{L}\mathscr{F})$ space is a Hausdorff LCTVS E which is an inductive limit of a sequence (E_i) of $(\mathscr{F})$ spaces by linear mappings u_i, such that the $u_i(E_i)$ generate E.

The considerations of Section 1 show that we can suppose the sequence (E_i) to be an increasing sequence of vector subspaces of E with given topologies T_i which make them into $(\mathscr{F})$ spaces, the identity mapping from E_i into E_{i+1} being continuous and E being the union of the E_i. Such a sequence is called a sequence of definition of the $(\mathscr{L}\mathscr{F})$ space. The most important examples of Section 2 are $(\mathscr{L}\mathscr{F})$ spaces (examples c), e), d) when M is countable at infinity, example f) when A is open or compact). An $(\mathscr{F})$ space is trivially of type $(\mathscr{LF})$. By the transitivity of inductive limits, a Hausdorff E space which is an inductive limit of a *sequence* of $(\mathscr{LF})$ spaces by linear mappings such that the union of their images generates E, is an $(\mathscr{LF})$ space; in particular, a quotient space of an $(\mathscr{LF})$ space by a closed vector subspace is an $(\mathscr{LF})$ space. It should be noted carefully that a closed vector subspace of an $(\mathscr{LF})$ space or even of a topological direct sum of a sequence of Hilbert spaces may not necessarily be of type $(\mathscr{LF})$. The spaces of type $(\mathscr{LF})$ are bornological and barrelled (Proposition 1, Corollary) but not necessarily complete even in the case of inductive limits of Banach spaces (Köthe's example, Section 4, Exercise 4). The more special properties of $(\mathscr{LF})$ spaces follow from the next theorem (where local convexity is not considered).

THEOREM 1   *Let E be a vector space with a Hausdorff topology, (E_i) a sequence of complete and metrisable TVS; let u_i be a continuous linear mapping from E_i into E for every i. Let u be a continuous linear mapping from a complete and metrisable TVS F into E, such that $u(F) \subset \bigcup u_i(E_i)$. Then there exists an index i such that $u(F) \subset u_i(E_i)$, and if u_i is bijective, we can write $u = u_i \circ v$, where v is a continuous linear mapping from F into E_i.*

Proof   Let H_i be the closed subspace of $F \times E_i$ consisting of pairs (x, y) such that $ux = u_iy$ and let $F_i = p_i(H_i) \subset F$ where p_i is the projection of $F \times E_i$ onto F; F_i is clearly the set of $x \in F$ such that $ux \in u_i(E_i)$.

The hypothesis states that $F = \cup F_i$, and (F being a Baire space) it follows that one of the spaces F_i is meagre. But by the Banach homomorphism theorem (Chapter 1, Section 14, Theorem 9, Corollary 3) this implies that P_i is a linear map from H_i onto F, i.e. $F_i = F$ and $u(F) \subset u_i(F_i)$.

If u_i is bijective we clearly have $u = u_i \circ v$ where v is a linear mapping from F into E_i which is continuous by the closed graph theorem (Chapter 1, Section 14, Theorem 10, Corollary) since it is continuous for the inverse image topology of E by u_i on E_i, which is a topology coarser than the given one and still Hausdorff.

COROLLARY 1   *Let E be an $(\mathscr{LF})$ space, (E_i) a sequence of definition of E, u a continuous linear mapping from a space F of type $(\mathscr{F})$ into E. Then there exists an i such that u is a continuous linear mapping from F into E_i (with its given $(\mathscr{F})$ space topology).*

This is an immediate particular case of Theorem 1. Applying this to the case $F = E_A$, a normed space defined by a bounded disk of E, we find

COROLLARY 2   *Let E be an $(\mathscr{LF})$ space, (E_i) a sequence of definition of E, A a bounded disk in E such that E_A is complete (for example a bounded and complete disk). Then A is contained and bounded in a space E_i.*

In particular, if E is quasi-complete, then the bounded subsets of E come from bounded subsets of E_i (the converse is false, see the space E of Exercise 4 above).

COROLLARY 3   *Let E be a vector space, let (E_i) (resp. (F_i)) be a sequence of $(\mathscr{F})$ spaces, let for every i, u_i (resp. v_i) be a linear mapping from E_i (resp. F_i) into E; we suppose that the images of the E_i (resp. F_i) generate E and that we can find on E a Hausdorff topology for which the u_i and the v_i are continuous. Then, on E, the inductive limit topology of the E_i by the u_i is identical to the inductive limit topology of the F_i by the v_i.*

In fact, we can reduce this to the case where the (E_i) (or F_j)) form a sequence of definition and we see by Theorem 1 that every E_i is contained in some F_j (with the identity mapping $E_i \to F_j$ continuous) and conversely; the result follows. We then see that in practice there cannot be more than one reasonable $(\mathscr{LF})$ topology on a vector space.

Using Theorem 1 the theorem of homomorphisms and the closed graph theorem can be generalized. In order to do this we introduce the notion of a *strictly bornological space*: it is a Hausdorff LCTVS which is the inductive limit of a family (of any cardinality) of *Banach* spaces (this is equivalent to saying that in E, every $\mathfrak{S}$-absorbing disk, where $\mathfrak{S}$ is the set of bounded disks in E such that E_A is complete, is a neighborhood of zero; compare with Section 2, Example b)). If E is Hausdorff and quasi-complete for E to be bornological or strictly bornological is the same thing (Section 2, Example b)).

In particular, an $(\mathscr{F})$ space is strictly bornological since it is bornological (Chapter 3, Section 4, Theorem 3). It is trivial that every inductive limit of strictly bornological spaces is strictly bornological (transitivity of inductive limits), in particular, on $(\mathscr{LF})$ space is strictly bornological. We have,

THEOREM 2   *Let E be a strictly bornological LCTVS, F a Hausdorff LCTVS which is the union of a sequence of images of spaces E_i of type $(\mathscr{F})$ by continuous linear mappings u_i (for example E and F of type $(\mathscr{LF})$).*

1) *Every continuous linear mapping from F onto E is a homomorphism.*

2) *Let u be a linear mapping from E into F. For the mapping u to be continuous it is sufficient that its graph be closed and even that it is not possible to find in E a sequence (x_i) tending to zero such that the sequence (ux_i) tends to a limit different from zero.*

Proof

1) We can clearly suppose that u is bijective (replace F by $F/\ker u$); we must show that the identity mapping from E with the given topology into E with the inductive limit topology T of the E_i by the $v_i = u \circ u_i$ (written E_T) is continuous. Since E is strictly bornological this means that for every Banach space H and every continuous linear mapping w from H into E, w is also a continuous mapping from H into E_T (Section 1, Proposition 2). But we can clearly assume that the union of the images of the E_i in E is identical to E and that the v_i are bijective, the result then follows from Theorem 1.

2) In order to verify the continuity of u we are reduced to the case where E is a Banach space. Let $H = E \times F$ the graph of u, let

$$H_i = H \cap (E \times E_i)$$

(we suppose that the E_i are subspaces of F whose union is F); the weakest hypothesis on 2 implies that H_i is a closed subspace of $E \times E_i$, i.e. it is an $\mathscr{F}$ space for the induced topology. H is the union of the H_i and satisfies the condition of the space F of Statement 1. Therefore, the projection of H onto E is an isomorphism, i.e. the inverse mapping is continuous, therefore u is continuous, u obtained by composition of the preceding mapping with the projection of H into F.

It seems we could considerably weaken the conditions on F, a question worth some research.

EXERCISE 1   Let E be an $(\mathscr{F})$ space, F an $(\mathscr{LF})$ space.

a) Show that if either E is normable, or F quasi-complete then the space of bounded linear mappings from E into F (Chapter 2, Section 18, Definition 14) can be given a uniquely determined $(\mathscr{LF})$ topology, finer than the topology of the bounded convergence (see Section 2, Example c)).

b) E and F being as in a), let G be an $(\mathscr{F})$ space, let u be a continuous bilinear mapping from $E \times G$ into F such that for every $z \in G$ the mapping $x \mapsto u(x, z)$ from E into F is bounded. Show that there exists a fixed neighborhood V of 0 in E, such that for every $z \in G$, V be transformed into a bounded subset of F by the mapping $x \mapsto u(x, z)$.

c) State results analogous to a) and b) when F is an $(\mathscr{F})$ space or more generally a strict inductive limit of a sequence of ($\mathscr{F}$) spaces and where the *bounded* mappings are replaced by *compact* mappings.

6 Products and direct sums of lines

PROPOSITION 8 *Let E be an LCTVS. The following conditions are equivalent:*

1) *The topology of E is the finest LCTVS topology on E (or equivalently, every linear mapping from E into an LCTVS is continuous).*

2) *Every linear form on E is continuous and the topology of E is a Mackey topology (Chapter 2, Section 13, Definition 12).*

3) *E is isomorphic to a topological direct sum of lines.*

The equivalence of the conditions in 1) is trivial, furthermore it is immediate that the finest locally convex topology on E is identical to the topology $\tau(E, E^*)$ (Mackey theorem), if on the other hand, (e_i) is a basis of E by which E is identified with k^I (k is the field of scalars), it is immediate from the definition that the finest locally convex topology on E is also the topological direct sum (which is the finest locally convex topology on E for which the mappings $\lambda \mapsto \lambda e_i$ are continuous, where *every* locally convex topology on E makes them continuous). This shows the equivalence of conditions 1), 2), 3).

PROPOSITION 9 *Let E be a vector space with the finest locally convex topology. E is bornological, complete, (therefore barrelled) and its bounded subsets are finite dimensional (a fortiori E is reflexive and even a ($\mathscr{M}$) space—see Chapter 2, Section 18, Definition 13). On the dual $E' = E^*$ of E the weak and strong topologies are identical and E^* is isomorphic to a topological product of lines.*

E is bornological and barrelled from the definitions (but also because of closure under inductive limits). The completion of E and the characterization of its bounded subsets is the result of Proposition 5, Section 3. The fact that the bounded subsets are finite dimensional is equivalent to the fact that on E' weak and strong topologies coincide. On the other hand, the weak dual of a topological direct sum is the product of the weak duals of the factors (Proposition 6), the result follows:

PROPOSITION 10 *Let E be a vector space with the finest locally convex topology. Then every vector subspace V of E is closed, the induced topology on V is also the finest locally convex topology on V, and every algebraic supplement of V is also a topological supplement (therefore V is a topological direct factor of E).*

Let W be an algebraic supplement of V, let p be the corresponding projection from E onto V; it is continuous when V has the finest locally convex topology T on V (Proposition 8, 1), it follows a fortiori that it is continuous from E into E, therefore that V and W are topological supplements and a fortiori V is closed; at the same time this proves that the induced topology of E onto V is identical to T (since the identity mapping of V onto V with one or the other topology is continuous).

COROLLARY   *Every quotient space of E is Hausdorff and its topology is the finest locally convex topology.*

In fact, this quotient space is by Proposition 10 isomorphic to a subspace W of E (supplementary to the given subspace V) hence the conclusion.

PROPOSITION 11   *Let E be a Hausdorff LCTVS. The following conditions are equivalent:*

1) *The topology of E is minimal among the Hausdorff locally convex topologies.*

2) $E = E'^*$.

3) *E is isomorphic to a topological vector product of lines.*

1) implies 2) since if we had $x \in E'^* \cap \complement E$ and if H is the kernel of the linear form x on E', the topology $\sigma(E, H)$ would be clearly Hausdorff, strictly coarser than $\sigma(E, E')$ (since its dual is H which is strictly smaller than E') therefore strictly coarser than the given topology.

Conversely, 2) implies that every equicontinuous subset of E', which will be weakly bounded for $\sigma(E', E'^*)$, is finite dimensional (since Proposition 10 contains the fact that for every vector space F, the bounded subsets for $\sigma(F, F^*)$, i.e. for the finest locally convex topology —which is the same by the Banach–Steinhaus–Mackey theorem—is finite dimensional. Therefore the topology of E is a weak topology. If T is a Hausdorff locally convex topology coarser than $\sigma(E, E')$ its dual must be E' (otherwise the topology would not be Hausdorff, every vector subspace of E' being weakly closed), therefore it is finer than the initial topology, therefore identical, which shows that the initial topology is minimal among the Hausdorff locally convex topologies on E. Finally, we know that E'^* with $\sigma(E'^*, E')$ is isomorphic to a topological product of lines (Proposition 9), and conversely such a space clearly satisfies 2).

L

PROPOSITION 12   *Let E be an LCTVS isomorphic to a topological vector product of lines. E is complete, barrelled, a ($\mathcal{M}$) space (a fortiori reflexive), whose strong dual is isomorphic to a topological direct sum of lines (considered in Proposition 9).*

Trivial (complete spaces (resp. barrelled, resp. of type ($\mathcal{M}$)) are closed under the formation of products; the strong dual of a topological vector product is the topological direct sum of strong duals). We do not know whether the space $\mathbf{R}^I$ is bornological (see Exercise 3).

PROPOSITION 13   *Let E be an LCTVS isomorphic to a topological vector product of lines, let V be a closed vector subspace of E; then, V admits a topological supplement, and the topology induced on V makes of it a space isomorphic to a topological vector product of lines.*

Consider the orthogonal V^0 of V in E' strong; we know (Proposition 10) that it admits a topological supplement N, and that V^0 and N are isomorphic to direct sums of lines. Then, the strong dual E of E' can be identified with the product of strong duals of V^0 and of N, these duals being isomorphic to topological vector products of lines; with this identification, V is the dual of N, hence the conclusion.

COROLLARY   *Every Hausdorff quotient space of E is also isomorphic to a topological vector product of lines.*

EXERCISE 1

a) Every continuous linear mapping from an LCTVS F into a space E isomorphic to a topological direct sum of lines is a homomorphism onto a closed subspace. If this mapping is bijective (or onto) it admits a left inverse (or right inverse) continuous linear mapping. In particular, if H is a vector subspace of F such that F/H is isomorphic to a direct sum of lines, then H admits a topological supplement.

b) Every continuous linear mapping from a space E isomorphic to a topological vector product of lines into a Hausdorff LCTVS F is a homomorphism onto a closed subspace. If this mapping is bijective (or onto) it admits a left inverse (or right inverse) continuous linear mapping. In particular, if H is a vector subspace of F isomorphic to a topological vector product of lines, then H admits a topological supplement.

EXERCISE 2   Let E be an LCTVS isomorphic to a topological direct sum of lines, F an LCTVS isomorphic to a topological vector product of lines; consider $H = E \oplus F$: spaces of this type are said to be linearly locally compact. Show that every closed vector subspace of H is also linearly locally compact and admits a topological supplement.

EXERCISE 3   Let I be an infinite set of indices.

a) Consider $\mathbf{R}^I$ and its dual $\mathbf{R}^{(I)}$. Show that every linear form on $\mathbf{R}^I$ bounded on the bounded subsets can be decomposed uniquely as $x' = y' + z'$, where $y' \in \mathbf{R}^{(I)}$ and where z' is zero on $\mathbf{R}^{(I)} \subset \mathbf{R}^I$.

b) Let z' be a linear form on $\mathbf{R}^I$ bounded on the bounded subsets and zero on $\mathbf{R}^{(I)}$. Show that we can find a finite partition of I into subsets J with the following property: for every decomposition of J into complementary sets J_1 and J_2, the restriction of z' to one of the two spaces $\mathbf{R}^{J_1}$ or $\mathbf{R}^{J_2}$ is zero (proceed by contradiction showing that we could otherwise find an infinite sequence $x^{(i)}$ of elements of E, with pairwise disjoint supports, and on which z' is not zero; we could suppose $\langle x^{(i)}, z' \rangle = 1$, which is absurd, since $x^{(i)}$ tends to zero in the sense of Mackey).

c) Let z' be a linear form on $\mathbf{R}^I$ bounded on the bounded subsets, zero on $\mathbf{R}^{(I)}$ and such that for every subset J of I the restriction of z' to $\mathbf{R}^J$ or to $\mathbf{R}^{(J)}$ is zero. Show that if z' is not zero then the subsets J of I such that the restriction of z' to $\mathbf{R}^J$ is not zero form an *ultrafilter* ϕ such that the intersection of a sequence of sets J_i belonging to ϕ still belongs to ϕ ("Ulam–Mackey ultrafilter") and that we would have

$$\langle x, z' \rangle = \lambda \lim x_i$$

for every $x = (x_i) \in \mathbf{R}^I$, where λ is a constant. Show that conversely for every Ulam–Mackey ultrafilter ϕ on I the preceding formula (with $\lambda = 1$) defines a linear form ε_ϕ on $\mathbf{R}^I$, bounded on the bounded subsets, zero on $\mathbf{R}^{(I)}$.

d) Show that the forms ε_ϕ thus constructed are linearly independent and that then every linear form on $\mathbf{R}^I$ bounded on the bounded subsets can be put uniquely in the form of a sum of an element of $\mathbf{R}^{(I)}$ and a linear combination of linear forms of type ε_ϕ.

e) The space $\mathbf{R}^I$ is bornological if and only if there does not exist any Ulam–Mackey ultrafilter on I. We then say that the cardinal of I is bornological. Show that if there exists a non-bornological cardinal then there exists a smallest non-bornological cardinal K, and, then a cardinal is bornological if and only if it is strictly inferior to K.

f) Show that a cardinal number which is the sum of a family of bornological cardinals, the cardinal of the family of indices being bornological, is bornological (use Chapter 3, Section 4, Exercise 2).

Ulam has shown that if α is bornological, 2^α is bornological; therefore, the cardinal K of e), if it exists, is a limit cardinal.

g) A cardinal C is *strongly inaccessible* if it is $> \aleph_0$, if it is not the sum of a family of strictly inferior cardinals, the family having a power strictly inferior to C, and if furthermore $\alpha < C$ implies $2^\alpha < C$ (which implies that C is a limit cardinal). The cardinal K if it exists, see e), is strongly inaccessible. We do not know of the existence of strongly inaccessible cardinals and we can almost surely add to the axioms of set theory the non-existence of strongly inaccessible cardinals without finding any contradiction. In such a system of axioms we would have the theorem: every space $\mathbf{R}^I$ is bornological (therefore—see Chapter 3, Section 4, Exercise—every product of bornological spaces is bornological).

h) Let z' be a linear form on $\mathbf{R}^I$ bounded on the bounded subsets, let E be a subset of $\mathbf{R}^I$ whose power is bornological. Show that the restriction of z' to E is continuous. There is an analogous statement for the convergent filters on $\mathbf{R}^I$ with a basis of bornological power.

In this order of ideas see also Bourbaki, *Integration*, Chapter IV, Section 4, Exercise 18.

PART 2   METRISABLE LCTVS

1 Preliminaries

Recall that a space $(\mathscr{F})$ is a metrisable and complete LCTVS. If E is a metrisable LCTVS, F a closed vector subspace, then E/F is metrisable, and if E is complete, so is E/F (Chapter 1, Section 4, Proposition 6). Clearly, a closed vector subspace of a space $(\mathscr{F})$ is a space $(\mathscr{F})$, as well as the topological vector product of a sequence of $(\mathscr{F})$ spaces.

A space E of type $(\mathscr{F})$ is a *Baire space*, a reason for its special properties. It is then a *barrelled* space, i.e. if F is an LCTVS, every subset of $L(E, F)$ bounded for pointwise convergence is equicontinuous (Banach–Steinhaus theorem; in fact, local convexity is irrelevant in this case, see Chapter 1, Section 15, Theorem 11). We conclude that if E and F are $(\mathscr{F})$ spaces, then every separately continuous bilinear mapping from $E \times F$ into an LCTVS G is continuous, and every set of bilinear mappings from $E \times F$ into G bounded for simple convergence is equicontinuous (Chapter 1, Section 15, Theorem 12). Among the properties using the fact that we are dealing with *complete* metrisable spaces we must point out the *closed graph theorem*, which in practice means that all linear mappings from an $(\mathscr{F})$ space which we shall find are continuous, and the *theorem of homomorphisms* which is a useful variant (Chapter 1, Section 14).

A locally convex metrisable space E (complete or not) is *bornological*, in particular its topology is the associated Mackey topology $\tau(E, E')$ (Chapter 3, Section 4, Theorem 3). In particular, if F is an LCTVS, the continuous or weakly continuous linear mappings from E into F are identical; similarly, there is identity between the homomorphisms and the weak homomorphisms (Chapter 2, Section 16, Proposition 28, Corollary 2 and Proposition 29, Corollary 3; notice that we use the fact that, the subspace $u(F)$ of E, being metrisable, has the Mackey topology).

We point out finally the following proposition of General Topology, which can be applied in particular to quotients of $(\mathscr{F})$ spaces:

PROPOSITION 1   *Let E be a complete metric topological space, $\mathscr{R}$ a separated and open equivalence relation in E. Then every compact subset of $E/\mathscr{R}$ is contained in the canonical image of a compact subset of E; every convergent sequence in $E/\mathscr{R}$ is the canonical image of a convergent sequence in E.*

Proof   The two parts of the proposition result from

COROLLARY 1   *Under the conditions of Proposition 1, let K be a totally discontinuous compact space, u a continuous mapping from K into $E/\mathscr{R}$. Then there exists a continuous mapping v from K into E such that*

$$u = \phi \circ v,$$

where ϕ is the canonical mapping from E onto $E/\mathscr{R}$.

The first part of the proposition follows since every compact subset A of $E/\mathscr{R}$ can be considered as the image of a compact K, totally discontinuous, by a continuous mapping from K onto A (we can take K to be the Čech–Stone compactification of A considered as a discrete set, or also—since A is metrisable—the tryadic Cantor set); similarly, the second part of the proposition follows from the lemma, considering a convergent sequence in $E/\mathscr{R}$ as the image of the compact space K formed by points $0, 1, \frac{1}{2}, \ldots, 1/n, \ldots$ by a continuous mapping from K into $E/\mathscr{R}$.

Proof of Corollary 1   We construct by induction an increasing sequence of finite partitions $V_1, \ldots, V_n, \ldots$ of K by sets simultaneously open and closed, and of mappings $U_{n,i} \to B_{n,i}$ assigning to every $U_{n,i} \in V_n$ an open ball of radius $\leqslant 1/n$ in E, such that $u(U_{n,i}) \subset \phi(B_{n,i})$ and that

$U_{n+1,i} \subset U_{n,j}$ implies $B_{n+1,i} \subset B_{n,j}$. The possibility is immediate because if the construction is made up to rank n, we consider for every $U_{n,i} \in V_n$ the covering formed by inverse images by u of images by ϕ of open balls of radii $< 1/n$ contained in $B_{n,i}$; since ϕ is an *open* mapping, we obtain an *open* covering of the totally discontinuous compact space $U_{n,i}$; therefore, there exists a finer finite covering defined by a partition of $U_{n,i}$ into sets simultaneously open and closed $U_{n+1,j}$. To each of these sets we can assign an open ball $B_{n+1,j}$ of radius $\leqslant 1/(n+1)$ in $B_{n,i}$ such that $u(U_{n+1,j}) \subset \phi(B_{n+1,j})$. This being done for all the $U_{n,i} \in V_n$, we obtain the covering $V_{n+1} = (U_{n+1,j})$ sought for, as well as the mapping $U_{n+1,j} \rightarrow B_{n+1,j}$. Let now $x \in K$, let $U_n(x)$ be the element of V_n containing x, let $B_n(x)$ be the associated ball in E_j we thus obtain a decreasing sequence of balls whose radii tend to zero, therefore (E complete) a limit point $v(x)$. Since $u(x) \in \phi(B_n(x))$ for every n, we conclude that $\phi(v(x)) = ux$. On the other hand, the oscillation of v in an element of V_n is at most $2/n$, therefore v is continuous.

We remark that in the case where $E/\mathscr{R}$ is the quotient space of an ($\mathscr{F}$) space by a closed vector subspace F, Proposition 1 is a particular case of a stronger result (Chapter 1, Section 14, Exercise 2).

EXERCISE   Let E be metrisable and compact, $\mathscr{R}$ a separated equivalence relation in E such that every convergent sequence in $E/\mathscr{R}$ is the image of a convergent sequence in E. Show that $\mathscr{R}$ is open.

2 Bounded subsets of a metrisable LCTVS

THEOREM 1   *Let E be a metrisable LCTVS.*

1) *For every sequence (A_i) of bounded subsets of E, there exists a sequence (λ_i) of numbers > 0 such that $\cup \lambda_i A_i$ is bounded; furthermore, there exists a closed and bounded disk A in E such that the A_i are bounded subsets of the normed space E_A.*

2) *Let A be a bounded subset of E. There exists a closed and bounded disk B of E such that $A \subset B$, and such that the topology and the uniform structure induced by the normed space E_B on A is identical to that induced by E.*

It is immediate that the two assertions of 1) are equivalent. Let (p_n) be a fundamental sequence of semi-norms in E, let

$$M_i^n = \sup_{x \in A_i} p_n(x),$$

we will then have, if $x \in \cup \lambda_i A_i$,

$$p_n(x) \leqslant \sup \lambda_i M_i^n;$$

it suffices to choose (λ_i) such that for every n we have

$$\sup_i \lambda_i M_i^n < +\infty,$$

or equivalently, such that for every n, (λ_i) is bounded from above by a multiple of the sequence $(1/M_i^n)_i$. It suffices in fact to choose

$$\lambda_i = \inf \left(\frac{1}{M_i^1}, \ldots, \frac{1}{M_i^n} \right).$$

In order to prove the second part we can suppose that A is a disk, is then sufficient for E_B to induce on A the same system of neighborhoods of 0 which E induces (Chapter 2, Section 14, Lemma), or for every $\lambda > 0$, an index n exists such that $A \cap V_n \subset \lambda B$ (where (V_n) is a fundamental sequence of neighborhoods of 0 in E). Now we have $A \subset \cap \lambda_i V_i$ where (λ_i) is some sequence of numbers > 0; let (μ_i) be a sequence of positive numbers such that $\lambda_i/\mu_i \to 0$; we contend that $B = \cap \mu_i V_i$ satisfies the requirement. In fact, we have $A \subset \lambda \mu_i V_i$ for i sufficiently large (that is for i such that $\lambda_i < \lambda \mu_i$), now, let n be such that V_n is contained in the intersection of the $\lambda \mu_i V_i$ for the other indices i, then we have $A \cap V_n \subset \lambda \mu_i V_i$ for *every* i, i.e. $A \cap V_n \subset \lambda B$. The two statements of the theorem give

COROLLARY 1 *Let E be a metrisable LCTVS, let (A_i) be a sequence of bounded subsets of E. Then there exists a closed and bounded disk A in E such that the A_i are bounded subsets of the normed space E_A and such that the latter induces on them the same topology and uniform structure as does E.*

COROLLARY 2 *Let A be a precompact (resp. compact, resp.weakly compact and convex) subset of the metrisable LCTVS E. Then there exists a closed and bounded disk B in E such that A is also a precompact (resp. compact, resp. weakly compact) subset of the normed space E_B.*

This follows trivially from Theorem 1, 2) in the case where A is precompact or compact. In the case of weak compactness we use the fact that for a *convex* subset A of a locally convex space $F(= E_B)$, the fact of weak compactness depends only on the topology induced by F (not F weak but F) on A (Chapter 2, Section 9, Exercise 2). We shall see in Chapter 5 that in a *complete* LCTVS, the closed convex hull of a weakly compact set is weakly compact, therefore if E is an $(\mathscr{F})$ space, it is useless to suppose A convex in the preceding corollary.

COROLLARY 3 *Let f be a continuous mapping from a locally compact, τ-compact space M into a metrisable LCTVS E. Then there exists a closed and bounded disk A in E such that f is a continuous mapping from M into the normed space E_A.*

The remark following Corollary 2 shows that Corollary 3 remains valid if we replace continuity by weak continuity.

EXERCISE 1 Show the analogue of Corollary 3 above for the continuous functions zero at infinity (it is not necessary to suppose M τ-compact). What do we obtain when M is the set of natural numbers, with the discrete topology?

EXERCISE 2 Let M be a locally compact space with a positive measure μ, where M is the union of a sequence of integrable sets. Let f be a measurable mapping from M into a metrisable LCTVS E. Show that f is almost everywhere equal to a measurable mapping from M into a normed space E_A, where A is a closed and bounded disk in E.

EXERCISE 3 Let $\lambda^n = (\lambda_i^n)_i$ be a sequence of sequences of positive numbers. Let E be a metrisable LCTVS, (x_i) a sequence in E such that for every $x' \in E'$, the sequences $(\lambda_i^n \langle x_i, x' \rangle)$ are bounded for every n. Show that there exists a closed and bounded disk A in E such that the x_i belong to the normed space E_A, and that the sequence of their norms $|| x_i ||_A$ is such that the sequences $(\lambda_i^n || x_i ||_A)_i$ are bounded for every n.

EXERCISE 4 Let E be a Hausdorff LCTVS. Show that the following conditions are equivalent:

a) For every compact linear mapping from a LCTVS F into E, the transpose is a compact linear mapping from E' strong into F' strong.

b) The same as a) but F being a Banach space.

c) For every compact disk A in E, there exists a closed and bounded disk $B \supset A$ such that A is a compact subset of the normed space E_B. This means even that the transpose of a compact linear mapping from an LCTVS F into E is a compact mapping from E'_c (E' with precompact convergence) into F' strong. Apply this result to the case where E is a metrisable space.

3 T_c Topology on the dual

Let E be an LCTVS, then on the equicontinuous subsets of E' (which are also uniformly equicontinuous), the topology T_c of uniform convergence on precompact sets T_c is identical to the weak topology.

Therefore on E' the topology T_c is a priori coarser than the finest topology T_0 on E' (locally convex or not) inducing on the equicontinuous subsets the weak topology (see Chapter 0, Section 1). We shall see that they are indentical if E is metrisable (this forms with the Hahn–Banach theorem, the Banach–Steinhaus theorem, the theorem of homomorphisms or the closed graph theorem, one of the deepest results of the theory, although less frequently used than the others):

THEOREM 2 (BANACH–DIEUDONNÉ) *Let E be a metrisable LCTVS. Then on E' the topology of precompact convergence is identical to the finest topology which induces on the equicontinuous subsets the weak topology.*

We first give the corollaries. The theorem is equivalent to

COROLLARY 1 *A subset H of E' is closed for the topology T_c if and only if for every weakly closed equicontinuous subset A of E', $A \cap H$ is weakly closed. Equivalently: A subset U of E' is open for T_c if and only if for every equicontinuous subset A of E', $U \cap A$ is relatively open in A with the weak topology.*

Or equivalently:

COROLLARY 2 *Let u be a mapping from E' into a topological space F. The mapping u is continuous for the T_c topology on E' if and only if its restriction to every equicontinuous subset of E' is weakly continuous.*

Corollaries 1 and 2 are, for example, useful when E is a metrisable space of type $(\mathcal{M})$, since then the topology T_c is the strong topology, important in itself. In the general case, the most important application is

COROLLARY 3 *Let E be an $(\mathcal{F})$ space, let H be a convex subset of E'. The subset H is weakly closed if and only if its intersection with every weakly closed equicontinuous subset of E' is weakly closed.*

Since the dual of E' for T_c is E (Mackey theorem), for a convex subset H of E' to say that H is closed for T_c is to say that it is weakly closed, and the conclusion follows from Corollary 1.

Furthermore, we shall show that the finest topology T_c which . . ., is identical to the topology of uniform convergence on the *sequences* of E converging to 0. From this we get:

COROLLARY 4 *Let E be a metrisable LCTVS. Then on E' the topologies of precompact convergence, compact convergence and uniform convergence on the sequences of E tending to 0 are identical. In other words, for every* *precompact subset K of E, there exists a sequence (x_i) in E tending to 0 such that K is contained in the closed disked hull of (x_i).*

Proof of Theorem 2   We shall show that for every subset U of E' which is open for the topology T_0, i.e. such that the intersection of $\complement\, U$ with every weakly closed equicontinuous subset of E' is weakly closed, and every $x' \in U$, there exists a neighborhood of x' for the uniform convergence on the sequences converging to 0, contained in U. We can clearly suppose $x = 0$ and we have to find a set K in E, the set of points of a sequence converging to 0, such that $K^0 \subset U$. Let (V_n) be a fundamental decreasing sequence of neighborhoods of 0 in E, and construct by induction a sequence $K_0, \ldots, K_n, \ldots$ of finite subsets of E with

$$(1)\begin{cases} K_n \subset V_n \text{ for } n \geqslant 1, \\ K_n'^{\,0} \cap (V_{n+1})^0 \subset U \text{ for } n \geqslant 0 \text{ (where } K_n' = \bigcup_{i < n} K_i). \end{cases}$$

Then $K = \cup K_n$ satisfies the required conditions (since E' is the union of polars of V_n). The construction of K_0 is immediate since it is only necessary that K_0 be a finite subset whose polar does not meet $V_1^0 \cap \complement\, U$, which is possible since this last set is weakly closed and does not contain 0. In order to construct K_{n+1} we must find a finite subset $A = K_{n+1}$ of V_{n+1} such that $(K_n' \cup A)^0$ does not meet $(V_{n+2})^0 \cap \complement\, U$. Now this last set is weakly compact and its intersections with the $(K_n' \cup A)^0$ form a filter base of weakly closed sets, whose intersection is empty since a point in the intersection belongs to

$$(K_n' \cup V_{n+1})^0 = K_n'^{\,0} \cap (V_{n+1})^0,$$

which is contained in U by the induction assumption. Then at least one of the sets

$$(K_n' \cup A)^0 \cap (V_{n+2})^0 \cap \complement\, U$$

is empty, and the result follows.

EXERCISE 1   Let E be an LCTVS. Consider the following conditions:

a) Every vector subspace of E' whose intersections with the weakly closed equicontinuous subsets are weakly closed, is weakly closed;

b) Every continuous linear mapping from E onto a barrelled LCTVS F is a homomorphism.

1) Show that a) implies b) and that if F is barrelled, b) implies a). Application: Deduce the theorem of homomorphisms for $(\mathscr{F})$ spaces from Theorem 2, Corollary 3.

2) Show that the properties a) and b) are closed under the formation of quotients by a closed vector subspace. Show that a) implies that E is complete (therefore that every quotient space of E by a closed vector subspace is complete).

3) Conclude that properties a) and b) are not necessarily true when E is the topological direct sum of a sequence of spaces isomorphic to c_0 or to l^1 (use Part 1, Section 4, Exercise 5). A fortiori Theorem 2 is not necessarily true for these direct sums.

4) In 1) we have seen that the theorem of homomorphisms is valid for a continuous linear mapping from a space F of type $(\mathscr{F})$ onto a barrelled LCTVS F. It is no longer valid if we suppose E barrelled, F of type $(\mathscr{F})$: let E be a normed space of codimension 1 in its completion $\hat{E}$ (therefore barrelled, see Chapter 3, Section 2, Exercise 7, 2)), D a line supplementary to E in $\hat{E}$, $F = \hat{E}/D$, u the bijective continuous linear mapping from E onto F induced by the canonical mapping from $\hat{E}$ onto F. The mapping u is not an isomorphism (since E is not complete).

EXERCISE 2   Let E be a barrelled LCTVS, u a linear mapping from E into a LCTVS F.

1) Let H be the subspace of F' formed by the y' such that $y' \circ u$ is continuous. Show that if A is a weakly bounded subset of F' contained in H, then its weak closure is contained in H. Conclude that if F satisfies condition a) of Exercise 1, (in particular if F is an $(\mathscr{F})$ space), then H is a *weakly closed* vector subspace of F'.

2) Conclude from 1) that if E is an $(\mathscr{F})$ space, u is continuous if we can find a total subset in F' formed by y''s such that each $y' \circ u$ is continuous. (It suffices to verify that u is weakly continuous.)

3) Let u be a linear mapping from a barrelled LCTVS E into a space F of type $(\mathscr{F})$. Show that u is continuous if we can find on F a Hausdorff locally convex topology coarser than that given on F for which u is continuous (immediate consequence of 2)).

4) The mapping u is continuous if every $y' \in F'$ is in the weak closure of a weakly bounded set of F' whose elements y' are such that $y' \circ u$ is continuous.

EXERCISE 3   Let E be an $(\mathscr{F})$ space, $(x_i)_{i \in I}$ a family of elements of E, H a total subset of E' such that for every family $\lambda = (\lambda_i)$ of scalars all equal to $+1$ or -1, we can find an element $x = u(\lambda)$ in E (evidently unique) such that

$$\langle x, x' \rangle = \sum \lambda_i \langle x_i, x' \rangle$$

for every $x' \in H$. Show that the map u can be extended to a continuous linear map from $l^\infty(I)$. (Let F be the subspace of $l^\infty(I)$ formed by linear combinations of $\lambda = (\lambda_i)$, extend u into a linear mapping u from F into E; show that u is continuous using Exercise 2, and the fact that F is barrelled, established in Chapter 3, Section 3, Exercise 7). What can be said about the summability of $(X_i)_{i \in I}$? (See Chapter 2, Section 18, Exercise 3).

THEOREM 3   *Let E and F be $(\mathscr{F})$ spaces, u a continuous linear mapping from E into F. The following conditions are equivalent:*

1) *u is a homomorphism.*

2) *u is a weak homomorphism.*

3) *$u(E)$ is closed.*

4) *u' is a weak homomorphism.*

5) *$u'(F')$ is weakly closed.*

6) *$u'(F')$ is strongly closed.*

It suffices in order to verify these conditions that we have

7) *u' is a strong homomorphism.*

This condition is also necessary if E and F are Banach spaces.

Proof   1) is equivalent to 2), since the topology of $u(E)$ is the Mackey topology (see Section 1). 1) implies 3) since every quotient space of E is complete, on the other hand 3) implies 1), which is the homomorphism theorem (Chapter 1, Section 14). Anyway, 3) is equivalent to 4) and 2) equivalent to 5) (Chapter 2, Section 16, Proposition 27). Therefore conditions 1) to 5) are equivalent and 5) clearly implies 6). We shall show that 6) also implies 5) by means of the following stronger result:

LEMMA   *Suppose that for every weakly closed disked equicontinuous subset A of E', $u'(F') \cap A$ is closed in the Banach space E'_A. Then $u'(F')$ is weakly closed (i.e. u is a homomorphism).*

(Notice that the hypothesis of the lemma means that $u'(F')$ is closed with respect to the *sequences* that converge to 0 in the *sense of Mackey*.) Applying Theorem 2 of Section 3, it suffices to show that for every A, $A \cap u'(F') = B$ is *weakly* closed because it is *weakly compact* as will be shown. Now B being a *closed* disk of the Banach space E'_A, $E'_B = (E'_A)_B$ is a Banach space, therefore a Baire space. On the other hand F' is the union of a sequence (C_n) of weakly compact disks. Then

$$B_n = A \cap u'(C_n)$$

is a disk of E'_B which is weakly compact in E' and a fortiori closed in E'_B, the union of the nB_n is E'_B. From this it follows that one of the B_n is a neighborhood of 0 in E'_B, then we shall have, if needed multiplying C_n by a scalar, $B \subset B_n = u'(C_n) \cap A$, whence $B = B_n$. Now B_n is weakly compact, so B is, hence the conclusion. (This lemma is due to G. Köthe.)

Thus, conditions 1) to 6) are equivalent. We now show that 7) implies 6). In fact, let G be the closure of $u(E)$, write $u = w \circ v$ where v is the mapping from E into G deduced from u and w the identity mapping from G into F. We then have $u' = v' \circ w'$, but since u' is a strong homomorphism so is v': if U is a strong neighborhood of 0 in G', since w' is a mapping *onto*, we have $U = w'(V)$, where $V = w'^{-1}(U)$ is a strong neighborhood of 0 in F', then $v'(U) = u'(V)$ and since by hypothesis $u'(V)$ is a strong neighborhood of 0 in $u'(F')$, $v'(U)$ is a strong neighborhood of 0 in $v'(G') = u'(F')$, so v' is a strong homomorphism. Since v' is bijective, v' is a strong isomorphism from G' into E', then, G' being complete, $u'(G')$ is a strongly complete subspace, hence strongly closed, of E'.

Finally, we have seen that (Chapter 2, Section 17, Proposition 32, Corollary 2) that if E and F are Banach spaces, 1) implies 7), which ends the proof of Theorem 3. Now we can state the converse of the result:

COROLLARY 1   *Let E and F be Banach spaces, u a continuous linear mapping from E into F. The mapping u is a homomorphism (or a metric homomorphism) from E into F if and only if u' is a homomorphism (or a metric homomorphism) from the Banach space F' into the Banach space E'.*

The case of a homomorphism follows from the equivalence of conditions 1) and 7) of Theorem 3. It remains to be shown that if u' is a metric homomorphism, so is u. Now u is a homomorphism, and by Chapter 2, Section 17, Proposition 32, in order to show that u is a *metric* homomorphism we can limit ourselves to the case where u is an *isomorphism* (topological) *onto*. This is trivial since the norm of a normed space is known once we know the unit ball of its dual. Particular cases:

COROLLARY 2   *Let E and F be Banach spaces, u a continuous linear mapping from E into F. The mapping u is an isomorphism (resp. a metric isomorphism, a resp. homomorphism onto, a resp. metric homomorphism*

*if and only if u′ is a homomorphism onto a weakly dense subspace of E′
(resp. a metric homomorphism onto a weakly dense subspace of E′,
resp. an isomorphism from F′ into E′, resp. a metric isomorphism from F′
into E′).*

Finally, using Proposition 1, of Section 1, we have (independently
of Theorem 3):

COROLLARY 3   *Let E and F be ($\mathscr{F}$) spaces, u a continuous linear mapping
from E into F. The mapping u is a homomorphism if and only if for every
sequence (y_i) in u(E) tending to 0, there exists a sequence (x_i) in E tending
to 0 such that $y_i = ux_i$ for every i.*

It should be noted that in general if u is a homomorphism from E
into F (E and F of type ($\mathscr{F}$)), $u′$ is not necessarily a strong homomorphism, in particular if u is an isomorphism from E into F, or a
homomorphism from E onto F. In other words (as we have already
pointed out in Chapter 2, Section 15), the strong dual of a closed subspace of a space E of type ($\mathscr{F}$) may not be identifiable with a quotient
of the strong dual of E, and the strong dual of a quotient space of E
may not be identifiable with a subspace of the strong dual of E (if we
wish the topologies to remain the same). Recall however that if u is an
isomorphism from a reflexive LCTVS into an LCTVS F, its transpose
is a strong homomorphism (Chapter 2, Section 15, Proposition 21, 2)).

EXERCISE 1   Let u be a homomorphism from a Banach space E into
an LCTVS F. Show that $u′$ is a strong homomorphism.

EXERCISE 2   Let P be the set of indices of derivation relative to $\mathbf{R}^n$.
Show that the linear mapping $f \mapsto (D^p f(0))_{p \in P}$ from $\mathscr{E}(\mathbf{R}^n)$ into $\mathbf{R}^p$ is a
homomorphism from the first *onto* the second (show that $u′(F′)$ is the
weakly closed subspace of $E′$ formed of distributions of support $\{0\}$).
Show that the kernel of this homomorphism has no topological supplement (if v were a right inverse of u, show that we could suppose
$v(y) \in \mathscr{E}(\mathbf{R}^n)$ has its support in a fixed compact set K; conclude by
contradiction, observing that the space of $f \in \mathscr{E}(\mathbf{R}^n)$ having its support
in K, admits a true continuous norm, while $\mathbf{R}^p$ does not).

PART 3   ($\mathscr{DF}$) SPACES

1 Generalities

DEFINITION 1   *A locally convex space H is a ($\mathscr{DF}$) space if it satisfies:*

  1) *H admits a fundamental sequence of bounded subsets.*

2) *If (U_i) is a sequence of closed and disked neighborhoods of 0 whose intersection U is bornivorous, then U is a neighborhood of 0.*

By polarity, condition 2) is equivalent to

2') *Every bounded subset M of the strong dual H' of H which is the union of a sequence of equicontinuous subsets is equicontinuous.*

Condition 2) is satisfied if H is quasi-barrelled (then H is of type $(\mathscr{D}\mathscr{F})$ if it admits a fundamental sequence of bounded sets), and in the general case it can be used to replace the requirement that a space is quasi-barrelled. A normed space satisfies condition 1) and is quasi-barrelled, therefore a $(\mathscr{D}\mathscr{F})$ space. Other examples will be studied in Section 4, but for the time being the most important one (which justifies the introduction of $(\mathscr{D}\mathscr{F})$ spaces) is given by

THEOREM 1 *The strong dual E' of a metrisable LCTVS E is of type $(\mathscr{D}\mathscr{F})$.*

In fact, the bounded subsets of E' are its equicontinuous subsets, and they admit a fundamental sequence (since E admits a fundamental sequence of neighborhoods of 0). Condition 2) remains to be verified; for this we notice that a fundamental system of neighborhoods of 0 in E' is formed of absorbing *weakly* closed disks (which are the polars of the bounded sets of E) and we must find such a V contained in U. Let (A_i) be a fundamental sequence of bounded disks of E' which we can suppose *weakly compact*; we construct by induction a sequence of *weakly* closed disked neighborhoods V_i of 0 and of scalars $\lambda_i > 0$ such that if the construction is done up to rank n, we have

$$\lambda_i A_i \subset \tfrac{1}{2}U; \ \lambda_i A_i \subset V_j; \ V_i \subset U_i$$

for $i, j \leqslant n$.

It suffices to set $V = \cap \, V_i$. The possibility of induction remains to be shown. We can find λ_{n+1} sufficiently small so that

$$\lambda_{n+1}A_{n+1} \subset V_i$$

for $i = 1, \ldots, n$ and

$$\lambda_{n+1}A_{n+1} \subset \tfrac{1}{2}U,$$

then
$$A = \Gamma_{i=1}^{n+1} (\lambda_i A_i)$$

is a *weakly compact* disk contained in $\tfrac{1}{2}U$. Let W be a weakly closed disked neighborhood of 0 contained in $\tfrac{1}{2}U_{n+1}$, then $V_{n+1} = A + W$ is a *weakly closed* disk contained in $\tfrac{1}{2}U + \tfrac{1}{2}U_{n+1}$ thus in U_{n+1} since U_{n+1} is a disk containing A, therefore the $\lambda_i A_i$ for $i = 1, \ldots, n + 1$.

PROPOSITION 1 *Let H be a $(\mathscr{D}\mathscr{F})$ space, E an LCTVS. Then every* *bounded subset of $L_b(H, E)$ which is the union of a sequence of equicontinuous subsets, is equicontinuous.*

This follows from Condition 2′ of Definition 1. A subset M of $L_b(H, E)$ is clearly bounded (resp. equicontinuous) if and only if for every equicontinuous subset A of E', the set $M'(A)$ the union of the $u'(A)$ when u runs through M is a bounded (resp. equicontinuous) subset of the strong dual of H.

COROLLARY 1 *Every bounded sequence in $L_b(H, E)$ and therefore every bounded subset M in which there exists a countable dense subset (it suffices that it be dense for pointwise convergence), is equicontinuous.*

In particular:

COROLLARY 2 *Let H be a complete $(\mathscr{DF})$ space, let (u_i) be a sequence of continuous linear mappings from H into E (E Hausdorff) converging pointwise to a limit $u(x)$. Then (u_i) is an equicontinuous sequence and u a continuous linear mapping and u_i tends to u uniformly on every compact set.*

In fact, since H is complete and (u_i) bounded for pointwise convergence, it is also bounded for bounded convergence, therefore equicontinuous (Corollary 1).

COROLLARY 3 *Let H be a $(\mathscr{DF})$ space, E an $(\mathscr{F})$ space, then $L_b(H, E)$ is an $(\mathscr{F})$ space.*

In fact, H admits a fundamental sequence of bounded subsets and since E is metrisable, $L_b(H, E)$ is metrisable. In order to verify completeness it suffices to see that every Cauchy sequence converges; for this, it suffices to verify that it converges pointwise, which follows from Corollary 2 (see also Section 5, Theorem 6, Corollary 1, for a more general and less obvious result), in particular:

COROLLARY 4 *The strong dual of a $(\mathscr{DF})$ space is of type $(\mathscr{F})$.*

By Theorem 1, it follows therefore that the bidual of a metrisable space E is an $(\mathscr{F})$ space (notice that as E is quasi-barrelled the natural topology of its bidual—uniform convergence on equicontinuous subsets —is identical to the topology of the strong dual of E strong).

EXERCISE 1 Let u be a weakly continuous linear mapping from a $(\mathscr{DF})$ space H into a *separable* LCTVS. Show that u is continuous (proceed by transposition, noticing that the equicontinuous subsets of E' are weakly separable).

EXERCISE 2   Let E be a reflexive non-separable Banach space, let H be the space E with the topology of uniform convergence on the separable bounded subsets of E'. Show that E is a reflexive $(\mathscr{DF})$ space whose topology is different from the Mackey topology. Show that the result of Exercise 1 is not necessarily true if E is not separable.

EXERCISE 3   Let E be a metrisable LCTVS. Show that every bounded and weakly bounded subset of E'' is contained in the weak closure of a bounded subset of E. In particular, every separable bounded subset of the completion of E is contained in the closure of a bounded subset of E.

EXERCISE 4   Let E be a metrisable LCTVS, $\mathfrak{G}$ a set of bounded subsets of E closed under unions. Show that E' with the $\mathfrak{G}$-topology is a $(\mathscr{DF})$ space. Application: the bidual of a $(\mathscr{DF})$ space H, with its natural topology (equicontinuous convergence) is a $(\mathscr{DF})$ space.

2 Bilinear mappings on the product of two $(\mathscr{DF})$ spaces

PROPOSITION 2   *Let H be a $(\mathscr{DF})$ space, (U_i) a sequence of neighborhoods of 0 in H. There exists a neighborhood U of 0 absorbed by every U_i (i.e. for every U_i some dilation of U_i contains U).*

By polarity, this means that for every sequence of equicontinuous subsets A_i of H', there exists a sequence of $\lambda_i > 0$ such that $\cup \lambda_i A_i$ is equicontinuous. Now H' being metrisable and the A_i bounded we can find the λ_i such that $\cup \lambda_i A_i$ is bounded (Part 1, Section 2, Theorem 1); this set is equicontinuous by Condition 2′ of Definition 1.

COROLLARY 1   *Every continuous linear mapping of a space H of type $(\mathscr{DF})$ into a metrisable LCTVS E is bounded (i.e. transforms some neighborhood of 0 into a bounded subset). Every equicontinuous set of linear mappings from H into E is equibounded (i.e. there exists a neighborhood U of 0 in H such that*

$$M(U) = \bigcup_{u \in M} u(U)$$

is a bounded subset of E).

It suffices to prove the second assertion. Let (V_n) be a fundamental sequence of neighborhoods of 0 in E, then for every n,

$$M^{-1}(V_n) = \bigcap_{u \in M} u^{-1}(V_n)$$

is a neighborhood U_n of 0 in H. If U is a neighborhood of 0 in H

M

I bsorbed by all the U_n, $M(U)$ is absorbed by all the V_n, i.e. it is bounded. an the same vein, we point out:

COROLLARY 2   *Let E be an ($\mathscr{F}$) space, H a ($\mathscr{DF}$) space. Every continuous linear mapping from E into H is bounded, every bounded set of continuous linear mappings from E into H is equibounded.*

It suffices to see the continuous linear mappings from E into H as bilinear forms on $E \times H'$ (product of two ($\mathscr{F}$) spaces) and to apply Chapter 1, Section 15, Theorem 12.

THEOREM 2   *Let H_1 and H_2 be ($\mathscr{DF}$) spaces, E an LCTVS, u a bilinear mapping from $H_1 \times H_2$ into E. The mapping u is continuous if and only if it is hypocontinuous (Chapter 3, Section 5, Definition). More generally a set M of bilinear mappings from $H_1 \times H_2$ is equicontinuous if and only if it is equi-hypocontinuous.*

Since M is equicontinuous (resp. equi-hypocontinuous) if and only if for every equicontinuous subset A of E' the set $M'(A)$ of bilinear forms $\langle u(x, y), z' \rangle$ on $H_1 \times H_2$ when u runs through M and z' runs through A is equicontinuous (resp. equi-hypocontinuous)—the verification is trivial; see Chapter 3, Section 5, Exercise 1—we are led back to the case where E is the field of scalars. But M can be identified with an *equicontinuous* set of linear forms from H_1 into the strong dual H_2 of H_2 (equi-hypocontinuity in H_2), and since H'_2 is an ($\mathscr{F}$) space, there exists a neighborhood U of 0 in H_1 such that $M(U)$ is a bounded subset of H'_2 (Proposition 2, Corollary 1). But H_1 is the union of a sequence of bounded subsets, thence also U with the A_i as bounded subsets, therefore

$$M(U) = \cup M(A_i).$$

The $M(A_i)$ are *equicontinuous* subsets of H'_2 (equi-hypocontinuity in H_1), thus $M(U)$ is a bounded subset of H'_2, the union of a sequence of equicontinuous subsets, therefore equicontinuous. This means that M is an equicontinuous set of bilinear forms.

COROLLARY 1   *Let H_1 and H_2 be barrelled ($\mathscr{DF}$) spaces; let E be an LCTVS, u a separately continuous bilinear form from $H_1 \times H_2$ into E; then u is continuous. More generally every set M of separately continuous bilinear mappings from $H_1 \times H_2$ into E which is pointwise bounded is equicontinuous.*

In the case of H_1 and H_2 barrelled, the hypothesis on M implies equi-hypocontinuity (Chapter 3, Section 5, Proposition 9, Corollary 1).

COROLLARY 2   *Let E_1 and E_2 be metrisable LCTVS, E an LCTVS, u a* *bilinear mapping from E_1' × E_2' into E separately weakly continuous. Then u is continuous for the product of the strong topologies. More generally, let M be a set of weakly separately continuous bilinear mappings from E_1' × E_2' into E, bounded for pointwise convergence, then M is equicontinuous for the product of the strong topologies.*

It suffices to consider the case where E is the field of scalars. The hypothesis on M implies that M is bounded for bi-bounded convergence (E_1' and E_2' are strongly complete), i.e. that for every bounded subset A of E_1', $M(A)$ is a bounded subset of E_2, therefore an equicontinuous subset of the dual of E_2' strong. This is a statement of the equi-hypocontinuity of M in E_1' strong; we see also that M is equi-hypocontinuous in E_2' strong, therefore, by Theorem 2, M is equicontinuous for the product of the strong topologies (E_1' and E_2' are ($\mathscr{DF}$) spaces by Theorem 1).

PROPOSITION 3 *Let H_1 and H_2 be ($\mathscr{DF}$) spaces, E an LCTVS, M a subset of the space $B_b(H_1, H_2; E)$ of continuous bilinear mappings from H_1 × H_2 into E with bi-bounded convergence. If M is bounded and is the union of a sequence of equicontinuous subsets, then M is equicontinuous.*

We may again assume that E is the field of scalars. M can be identified with a bounded set in $L_b(H_1', H_2')$ (where H_2 has the strong topology), and is the union of a sequence of equicontinuous subsets, therefore an equicontinuous subset of $L_b(H_1, H_2')$ (Proposition 1). Also, M defines an equicontinuous set of linear mappings from H_2 into H_1'. A fortiori, M is equihypocontinuous as a set of bilinear forms, therefore equicontinuous by Theorem 2.—The statements of the corollaries of Proposition 1 are left to the reader.

EXERCISE 1 Extend the preceding results to multilinear mappings on products of ($\mathscr{DF}$) spaces.

EXERCISE 2 Let E be a non-quasi-barrelled LCTVS. Show that there exists a normed space F (assumed complete if the strong dual of E is assumed complete), and a separately continuous bilinear form on $E \times F$ which is not continuous (choose in E' a closed and strongly bounded disk A not equicontinuous and set $F = E_A$). Conclude that in Corollary 1 of Theorem 2 we cannot restrict ourselves to the supposition that H_1 or H_2 is barrelled.

EXERCISE 3 An LCTVS which is simultaneously metrisable and of type ($\mathscr{DF}$) is normable (apply Proposition 2, Corollary 1).

3 Stability properties

PROPOSITION 4   *Let E be an* LCTVS, *F a vector subspace.*

1) *If F is a* $(\mathscr{DF})$ *space, then its strong dual can be identified with the quotient of the strong dual E' of E by the orthogonal F^0 of F.*

2) *If E is a* $(\mathscr{DF})$ *space, then E/F is a* $(\mathscr{DF})$ *space, and its strong dual can be identified with the subspace F^0 of the strong dual E' of E.*

In 1) we must show that the natural mapping

$$F'_b \rightarrow E'_b/F^0$$

is continuous; how F'_b is metrisable therefore bornological, so it suffices to verify that its sequences that converge to zero are transformed into bounded sequences. As F is a $(\mathscr{DF})$ space, a strongly bounded sequence of the dual is equicontinuous therefore comes from an equicontinuous (a fortiori strongly bounded) sequence of E' whose image in E'_b/F is therefore bounded. We show in the same way in 2) that the natural mapping from F^0 (with the metrisable topology induced by E'_b) into the strong dual of E/F transforms sequences converging to zero into bounded sequences, and is thus continuous, therefore an isomorphism. This is equivalent to saying that every bounded subset of E/F is contained in the closure of the canonical image of a bounded subset of E, from which it follows that a fundamental *sequence* of bounded sets in E/F exists (there exists one in the $(\mathscr{DF})$ space E). Finally, we trivially verify the second condition on the $(\mathscr{DF})$ spaces in E/F by the fact that it is verified in E.

COROLLARY 1   *Let E be an* LCTVS, *F a* $(\mathscr{DF})$ *vector subspace, then every bounded subset of the closure of F is contained in the closure of a bounded subset of F.*

We can assume F to be dense in E, the duals of E and F are then algebraically identical, and Proposition 4, 1) says that this identification respects the topologies which is exactly the meaning of the corollary. In particular:

COROLLARY 2   *Let F be a* $(\mathscr{DF})$ *space. The space F is complete if and only if it is quasi-complete. In particular, if F is reflexive it is complete.*

We point out that a closed vector subspace of a $(\mathscr{DF})$ space (even of type $(\mathscr{M})$) is not always of type $(\mathscr{DF})$. However, we verify trivially that the product of a *finite* number of $(\mathscr{DF})$ spaces is a $(\mathscr{DF})$ space. If E is a $(\mathscr{DF})$ space, its strong dual E' is an $(\mathscr{F})$ space, therefore its bidual with the topology of the strong dual of E' is a $(\mathscr{D}\mathscr{F})$ space (Theorem 1); E'' is still a $(\mathscr{D}\mathscr{F})$ space for its "natural" topology (Section 1, Exercise 4). Finally, we have

PROPOSITION 5   *Let E be an LCTVS inductive limit of a sequence of $(\mathscr{D}\mathscr{F})$ spaces E_i by linear mappings u_i such that the union of images of E_i generates E. Then E is a $(\mathscr{D}\mathscr{F})$ space, the strong topology of its dual is the coarsest for which the transposed mappings u_i' of E' into the strong duals E_i', are continuous; the bounded subsets of E are those contained in the closure of the sum of a finite number of images of bounded subsets of the E_i by the u_i.*

The two last assertions are equivalent and we must verify only that the identity mapping of E' with the coarsest topology T, for which the u_i' into E' strong are continuous, is continuous. Since T is metrisable it suffices to verify that a sequence that converges to 0 for T is strongly bounded, and it is even equicontinuous, since its image by every u_i' is a strongly bounded sequence therefore equicontinuous of E_i'. We have shown at the same time that E admits a fundamental *sequence* of bounded subsets. Finally, the second condition on $(\mathscr{D}\mathscr{F})$ spaces is trivially verified from the fact that it is valid on each E_i'. We could also prove Proposition 5 for the topological direct sums of a sequence of $(\mathscr{D}\mathscr{F})$ spaces by means of Proposition 4, 2).

COROLLARY 1   *If the E_i are reflexive $(\mathscr{D}\mathscr{F})$ spaces (resp. of type $(\mathscr{M})$) and E is Hausdorff, then E is reflexive (resp. of type $(\mathscr{M})$).*
We verify that its bounded subsets are relatively compact (resp. weakly relatively compact).

COROLLARY 2   *Let (E_i) be a sequence of LCTVS, let for every i u_i be a linear mapping from E_i into a vector space E and v_i a linear mapping from E_i into E_{i+1} such that $u_i = u_{i+1} \circ v_i$. Suppose E identical to the union of the images of the E_i and equip it with the inductive limit topology. If the mappings v_i are bounded then E is a quasi-barrelled and hornological $(\mathscr{D}\mathscr{F})$ space; if the mappings v_i are weakly compact (resp. compact) and E is Hausdorff then E is reflexive (resp. of type $(\mathscr{M})$) barrelled and bornological.*

Let for every i U_i be a disked neighborhood of 0 in E_i whose image by v_i is bounded, let F_i be E_i with the semi-norm gauge of U_i; we verify trivially that the topology of E is also the inductive limit topology of the semi-normed spaces F_i by the u_i. It is therefore a $(\mathscr{D}\mathscr{F})$ space quasi-barrelled and bornological since the F_i are, and furthermore every bounded subset of E is contained in the closure of the dilation of a set $u_i(U_i)$ (it is useless to choose finite sums since we already have a transitive system). Now $u_i(U_i) = u_{i+1}(A_{i+1})$ where $A_{i+1} = v_i(U_i)$. If the v_i are compact (resp. weakly compact), and if E is Hausdorff, then the $u_{i+1}(A_{i+1})$ are relatively compact (resp. weakly relatively compact) subsets of E which will then be of type ($\mathscr{M}$) (resp. reflexive). A fortiori it will be complete (Proposition 4, Corollary 2), thence barrelled since it is quasi-barrelled.

EXERCISE 1   With the hypothesis of Proposition 5, Corollary 2 (the v_i being *bounded* mappings) show that if the E_i are quasi-complete or the v_i weakly compact, then E is an ($\mathscr{LF}$) space (therefore *barrelled*).

EXERCISE 2   Let E be a ($\mathscr{DF}$) space. Show that its completion is ($\mathscr{DF}$). The space E is quasi-barrelled if and only if its completion is barrelled.

EXERCISE 3   Show that the *bornological* ($\mathscr{DF}$) spaces are exactly the inductive limits of a sequence of normed spaces whose images generate the space. A complete ($\mathscr{DF}$) space is bornological if and only if it is an ($\mathscr{LF}$) space.

EXAMPLES

a) Since the strong duals of ($\mathscr{F}$) spaces are of type ($\mathscr{DF}$), the distribution spaces $\mathscr{E}'$, $\mathscr{D}_{L^p}$, etc. (see L. Schwartz, *Theory of Distributions*) are ($\mathscr{DF}$) spaces.

Other important ($\mathscr{DF}$) spaces are often defined as inductive limits:

b) Let K be a compact set of $\mathbf{C}^n$, let $H(K)$ be the space of holomorphic functions defined in a neighborhood of K, with the inductive limit topology of $H(U)$ spaces, where U runs through the open neighborhoods of K (Part 1, Section 2, Example f)). Here it suffices that U runs through a fundamental sequence (U_n) of neighborhoods of K and we can suppose U_{n+1} relatively compact in U_n. Then the canonical mapping v_n from $H(U_{n+1})$ into $H(U_n)$ is *compact* (Montel theorem), therefore we have the conditions of Proposition 5, Corollary 2: $H(K)$ is a complete ($\mathscr{DF}$) space, barrelled, bornological and of type ($\mathscr{M}$).

c) Let (ϕ_n) be an increasing sequence of positive continuous functions on a locally compact space M, let for every n, E_n be the space of continuous functions f on M bounded above for the absolute value by a multiple of ϕ_n, with the natural norm assigning to f the smallest m such that $|f| \leqslant m\phi_n$, for which it is a Banach space (immediate verification). Then the space E of continuous functions on M bounded above by a multiple of one ϕ_n, is the union of an increasing sequence of E_n and it will have the corresponding inductive limit topology for which it will be a $(\mathscr{DF})$ space (Proposition 5) bornological and barrelled. If for example M is the union of a sequence of compact sets M_n and if we take an increasing sequence of positive continuous functions with compact support ϕ_n such that ϕ_n is 1 on K_n, E will be the space of continuous functions with compact support on M (Part 1, Section 2, Example d)). If $M = \mathbf{R}^p$ and if $\phi_n = (1 + r^2)^n$ (r is the distance to the origin), we find that E is a space of slowly increasing continuous functions (i.e. bounded above by a polynomial).

d) Variants are possible, supposing for example the ϕ_n defined on $\mathbf{C}^p$ and letting H_n be the subspace of E_n formed by holomorphic functions. It is a closed subspace (the uniform limit on a compact set of a holomorphic function is holomorphic by Weierstrass theorem), and the inductive limit of the H_n (space of holomorphic functions bounded above by a multiple of a ϕ_n) has thus the topology of a bornological and barrelled $(\mathscr{DF})$ space. Choosing for example $\phi_n = \exp nr$ (r distance to the origin) we find the space of entire functions of exponential type. Choosing $\phi_n = \exp r^n$ we find the space of entire functions of finite order, etc.

e) An open question is the set of conditions for a space $L_b(E, H)$, with E of type $(\mathscr{F})$ and H of type $(\mathscr{DF})$, to be of type $(\mathscr{DF})$ (we know that $L_b(E, H)$ will have a fundamental *sequence* of bounded subsets, an immediate consequence of Section 2, Proposition 2, Corollary 2). In particular, when H is the strong dual of an $(\mathscr{F})$ space F the problem is to know whether the space $B(E, F)$ of continuous bilinear mappings on the product of two $(\mathscr{F})$ spaces with the bi-bounded topology is a $(\mathscr{DF})$ space.

EXERCISE   Show that the space of entire function of exponential type, or of finite order on $\mathbf{C}^p$ is a $(\mathscr{DF})$ space of type $(\mathscr{M})$ (use Proposition 5, Corollary 2). Show that the first of these two spaces is isomorphic to the space $H(\{0\})$ of functions holomorphic in the neighborhood of the origin (represent a function holomorphic at the origin by the sequence of its Taylor coefficients).

4 Complementary results

As the title implies, in this section we will give some further results on $(\mathscr{DF})$ spaces but shall omit proofs (Theorem 3, Theorem 4, Lemma). The reader can proceed independently or look at Grothendieck: *Sur les espaces* $(\mathscr{F})$ *et* $(\mathscr{DF})$ in *Summa Brasiliensis Mathematicae*, where some varied counterexamples relative to ($\mathscr{F}$) and ($\mathscr{DF}$) spaces and some open questions are to be found.

THEOREM 3   *Let H be a ($\mathscr{DF}$) space, U a disk in H. The disk U is a neighborhood of 0 if and only if it induces on every bounded set a neighborhood of 0.*

We verify that this is equivalent to:

COROLLARY 1   *Let u be a linear mapping from H into an LCTVS E. The mapping u is continuous if and only if its restrictions to the bounded subsets of H are continuous.*

The analogous statement for the characterization of *equicontinuous* sets of linear mappings from H into E is also valid (same proof); this allows us to improve Theorem 2 of Section 2. Another application:

COROLLARY 2   *Let H be a ($\mathscr{DF}$) space, E a complete Hausdorff LCTVS, then $L_b(H, E)$ is complete.*

Every limit, for bounded convergence, of continuous linear mappings will have continuous restrictions to the bounded subsets and will be continuous by Corollary 1; then $L_b(H, E)$ is a *closed* subspace of the complete space of all mappings from H into E with bounded convergence, therefore complete. This result generalizes Proposition 1, Corollary 3. Compare with Chapter 3, Proposition 2.

THEOREM 4   *Let H be a ($\mathscr{DF}$) space, M a separable subset of H. Then, on M, the given topology T of H and the topology T_0 of uniform convergence on the strongly bounded subsets of H are identical.*

Since $T = T_0$ if and only if H is quasi-barrelled, we have:

COROLLARY 1   *A separable ($\mathscr{DF}$) space is quasi-barrelled.*

Another immediate consequence:

COROLLARY 2   *In H, the sequences that converge for T or for T_0 are identical. Equivalently, a mapping from a metrisable topological space into H is continuous for T if and only if it is continuous for T_0.*

Using Theorem 3 we have:

THEOREM 5   *A ($\mathscr{DF}$) space whose bounded subsets are metrisable is quasi-barrelled.*

In order to verify that the identity mapping from H with T into H

with T_0 is continuous, it suffices (Theorem 3, Corollary 1) to verify that its restriction to every bounded subset is continuous, by Corollary 2 of Theorem 4.

We can also prove:

LEMMA *Let (A_i) be an increasing sequence of bounded disks in the space H of type $(\mathscr{DF})$ such that their homothetics form a fundamental sequence of bounded subsets of H. Let $U = \cup A_i$, then the closure of U is identical to the unit ball associated with the gauge of U (i.e. to the union of the closures of segments intersected by U on the real lines passing through the origin).*

Using an argument of weak compactness, we obtain;

THEOREM 6 *Let E be a metrisable LCTVS, then every bornivorous disk. U in the strong dual E' contains a closed and bornivorous disk.*

By polarity, we obtain:

COROLLARY 1 *Every set of linear forms on E' uniformly bounded on every bounded subset, is contained in the weak closure, in the algebraic dual of E', of a bounded subset of E''.*

The most important particular case is

COROLLARY 2 *Let E be a metrisable LCTVS. Its strong dual E' is bornological if and only if it is quasi-barrelled (or barrelled, which amounts to the same because the space is complete).*

E' strong is quasi-barrelled as can be verified by Theorem 5. This is trivially true if E is reflexive, therefore:

COROLLARY 3 *The strong dual of a reflexive $(\mathscr{F})$ space is bornological.*

The fact of a strong dual being bornological can be useful in duality theorems such as the following:

PROPOSITION 6 *Let E be an LCTVS, F a quasi-barrelled vector subspace whose strong dual is bornological. This strong dual can be identified with the quotient of the strong dual E' by the subspace F^0 orthogonal to F.*

In order to verify that the canonical mapping $F'_b \rightarrow E'_b/F_0$ is continuous it suffices to verify that a bounded subset is transformed into a bounded subset, since F'_b is bornological; but F being quasi-barrelled, a bounded subset of F' strong is equicontinuous, therefore comes from an equicontinuous subset of E', a fortiori strongly bounded, whose canonical image in E'_b/F^0 is consequently bounded.

EXERCISE   Deduce from Theorem 4 that the theorem remains valid if we replace T and T_0 by the associated weak topologies. Give the corresponding analogue to Corollary 2 of Theorem 4. Show that the subsets of H which are compact for the weak topology associated with T or with T_0 are identical (see the case where H is complete and use Eberlein's theorem, to be proved in Chapter 5: In a complete locally convex Hausdorff space a subset is weakly relatively compact if and only if every sequence extracted from it admits a cluster point for the weak topology. Conclude that every linear form on H continuous for T_0 belongs to the completion of H' for $\tau(H', H)$; show that if H is complete, every linear form on H belonging to the completion of H' for $\tau(H', H)$ is bounded on the bounded subsets and that, conversely, if H is the strong dual of a metrisable LCTVS, every linear form on H bounded on the bounded sets, is continuous for T_0.

PART 4   QUASI-NORMABLE SPACES AND SCHWARTZ SPACES

1 Definition of quasi-normable spaces

DEFINITION 1   *A locally convex space E is quasi-normable if for every equicontinuous disk A in E' there exists an equicontinuous disk $B \supset A$ such that on A the topologies induced by E' strong and by E'_B are the same.*

By Chapter 2, Section 14, this is equivalent to saying that the uniform structures on A induced by E' strong or by E'_B are identical; or that the two topologies in question induce on A the same system of neighborhoods of 0. This means that for every $\lambda > 0$, there exists a strong neighborhood W of 0 in E' that we can suppose disked and weakly closed such that $A \cap W \subset \lambda B$. Supposing A and B weakly closed and letting their polars be U and V, letting the polar of W be M the condition of quasi-normability becomes: for every disked and closed neighborhood U of 0 in E, there exists another one V such that for every $\lambda > 0$ we can find a closed and bounded disk M such that $V \subset \lambda \bar{\Gamma}(U \cup M)$. Changing λM into M (change of notation) and noticing that

$$\tfrac{1}{2}(U + M) \subset \Gamma(U \cup M) \subset 2(U + M)$$

(the second inclusion is true because U is a neighborhood of 0), we verify that we can write the following instead of the inclusion written above:

$$(1) \quad V \subset \lambda U + M.$$

Thus, E is quasi-normable if and only if for every neighborhood U of 0 there exists a neighborhood V of 0 such that for every $\lambda > 0$ there exists a bounded set M such that we have (1): this is the form under which we shall verify quasi-normability in concrete cases.

A quasi-barrelled $(\mathscr{DF})$ space is quasi-normable: this is an immediate consequence of the definition and of Part 2, Section 2, Theorem 1, 2). An $(\mathscr{F})$ space even if it is of type $(\mathscr{M})$ (Chapter 2, Section 18, Definition 13) may not be quasi-normable, however, the $(\mathscr{M})$ spaces we find in practice are quasi-normable. The reader may verify that quasi-normable spaces are closed under the formation of quotients and topological direct sums, furthermore, the inductive limit of a sequence of quasi-normable spaces is quasi-normable (it is the quotient of the topological direct sum); the same is true for the topological vector product. (It suffices to use the definition; only the case of a direct sum calls for a proof, in which we proceed as in Part 2, Section 2, Theorem 1, 2)). A vector subspace of a quasi-normable space is not in general quasi-normable since every $(\mathscr{F})$ space is isomorphic to a subspace of the product of a sequence of Banach spaces (this product is quasi-normable by the preceding discussion) and we know there exist non-quasi-normable $(\mathscr{F})$ spaces.

We point out that from the definition it follows that if E is quasi-normable, then the equicontinuous subsets of its strong dual are metrisable. In the case where E is of type $(\mathscr{F})$, therefore E' strong of type $(\mathscr{DF})$ (Part 3, Theorem 1), we see with the aid of Part 3, Section 5, Theorem 5 and Theorem 6, Corollary 2, that E' strong is a *bornological* $(\mathscr{DF})$ space (if E is quasi-normable).

EXERCISE 1   Show that the bidual of a quasi-normable space is quasi-normable (use the condition stated in Formula (1)).

EXERCISE 2   Let E be an LCTVS. We say that E satisfies the condition of Mackey convergence if every sequence tending to 0 in E tends to 0 in the sense of Mackey (Chapter 3, Section 4, Definition 4), and that E satisfies the strict condition of Mackey convergence if for every bounded disk A in E, there exists a bounded disk $B \supset A$ such that on A, the topology induced by E or by E_B is the same (Example: metrisable spaces, by Part 2, Section 2, Theorem 1, 2)).

1) Let E be a quasi-barrelled LCTVS, then E is quasi-normable if and only if its strong dual satisfies the strict condition of Mackey convergence.

2) A subspace of a space which satisfies one of the two conditions of

Mackey convergence, the topological vector product of a sequence of spaces or the topological direct sum of a family of spaces which satisfy one of the two properties, satisfies also the same property.

3) Let E be an LCTVS admitting a fundamental sequence of bounded sets

a) If E satisfies the Mackey convergence condition, show that for every filter ϕ with a *countable base* on E converging to a limit $x \in E$, we can find a closed and bounded disk A such that $A \in \phi$ and such that the trace of ϕ on A tends to x in the normed space E_A; in particular, every *metrisable* bounded disk B in E is contained in a bounded disk A such that on B the topology induced by E or by E_A is the same.

b) Conclude from this that E satisfies the *strict* Mackey convergence condition if and only if its bounded subsets are metrisable and E satisfies the Mackey convergence condition for the sequences.

2 Lifting of strongly convergent sequences of linear forms on a subspace

An immediate consequence of Definition 1 is

PROPOSITION 1 *Let E be a quasi-normable LCTVS, (x_i') an equicontinuous sequence in E' tending strongly to a limit x'. Then there exists an equicontinuous disk A in E' such that x_i' tends to x' in the normed space E_A'; equivalently (letting $V = A^0$) there exists a neighborhood V of 0 in E such that x_i' tends to x' uniformly on V.*

If we suppose $x' = 0$ this means also that there exists a sequence of $\lambda_i > 0$ tending to 0, such that we have $x_i' \in \lambda_i A$. If we suppose that E is a topological vector subspace of an LCTVS F, we know (Hahn–Banach) that an equicontinuous subset A of E' is the canonical image of an equicontinuous subset B of F', therefore, under the preceding conditions, we can find for every i an extension y_i' of x_i' to F such that $y_i' \in \lambda_i B'$. Therefore:

THEOREM 1 *Let F be an LCTVS, E a vector subspace, (x_i') an equicontinuous strongly convergent sequence in E'. Then, if E is quasinormable, we can find extensions y_i' of the x_i' to F, such that (y_i') is a strongly convergent equicontinuous sequence in F'.*

Theorem 1 is often applied in the form of the

COROLLARY *Let E be a quasi-normable LCTVS whose topology is the coarsest for which certain linear mappings u_i from E into LCTVS E_i are continuous. Then the equicontinuous and strongly convergent sequences in*

E' can be obtained by taking finite sums of sequences that result from composing with u_i the elements of an equicontinuous and strongly convergent sequence in a space E_i'.

We may reduce this to the case where E is Hausdorff so that it can be identified with a topological vector subspace of the topological vector product of the E_i. We then apply Theorem 1 noticing that the strong dual of the product of the E_i can be identified with the topological direct sum of the strong duals of the E_i (Part 1, Section 4, Proposition 7), and that its equicontinuous subsets are those contained in the sum of a finite number of equicontinuous subsets of spaces E_i' (Chapter 2, Section 15, Proposition 22, Corollary 1).

We remark incidentally that Theorem 1 and Corollary 1 could also be stated for strongly convergent *filters* on an equicontinuous subset of the dual of E. If we suppose conversely that E is an LCTVS such that regardless of its inmersion into an LCTVS F, the statement thus reinforced of Theorem 1 is valid, then F is quasi-normable; we see this immediately by choosing E to be a quasi-normable LCTVS, for example a product of Banach spaces, which is always possible.

EXERCISE 1   Let E be a *separable* LCTVS, F a vector subspace, (x_i') an equicontinuous and weakly convergent sequence in F'; show that we can find extensions y_i' of the x_i' to E such that (y_i') is an equicontinuous and weakly convergent sequence in E'. Show that even if E is a Banach space, it is necessary in all of the preceding argument that E is separable (see Chapter 3, Section 7, Exercise 2, d)).

3 Quasi-normability and compactness

THEOREM 2   *Let E be a quasi-normable LCTVS, A an equicontinuous disk in E' which is compact for $\sigma(E', E'')$ (resp. for the strong topology), then there exists a weakly closed equicontinuous disk B in E' such that A is a weakly compact subset (resp. compact) of the Banach space E_B'.*

Choose B such that on A the topology induced by E' strong or by E_B' is the same. If A is strongly compact, it is compact also in E_B'. Similarly, if B is compact for the weak topology $\sigma(E', E'')$ associated with the strong topology, B is weakly compact in the Banach space E_B', since for a disked subset A of an LCTVS F (E_B' in this case) the fact of being weakly compact depends only on the topology induced by F on A (see Chapter 2, Section 9, Exercise 2).

COROLLARY   *Let u be a continuous linear mapping from a quasi-normable*

LCTVS E *into a Banach space F, transforming bounded subsets into weakly relatively compact subsets (resp. relatively compact). Then u is weakly compact (resp. compact).*

Let A be the image of the unit ball of F' by u', A is compact for $\sigma(E', E'')$ resp. for the strong topology, by Chapter 2, Section 18, Theorem 12 and Theorem 13. Let B be as in the statement of Theorem 2, let $U = B^0$; since $u(U)$ is clearly bounded (since $u(A^0)$ is contained in the unit ball of F) u is a continuous linear mapping from the space E with the semi-norm gauge of U. The dual of this semi-normed space is clearly E'_B and the transpose of u considered as a mapping from E_U into F is still u' considered as a mapping from F' into E'_B. Since this last mapping is weakly compact (resp. compact) so is u considered as a mapping from E_U (same reference), therefore $u(U)$ is weakly relatively compact (resp. relatively compact). In particular, we obtain:

COROLLARY 2   *Let E be a quasi-normable LCTVS. If E is reflexive resp. if its bounded subsets are precompact, a fortiori if E is of type $(\mathcal{M})$), then every continuous linear mapping from E into a Banach space F is weakly compact (resp. compact).*

This last property may not be true if E is not quasi-normable even if it is of type $(\mathcal{F})$ and $(\mathcal{M})$. We can find a space E of type $(\mathcal{F})$ and $(\mathcal{M})$ admitting a quotient which is a non-reflexive Banach space, then the canonical mapping from E onto its quotient is not even weakly compact! (At the same time we see that there exist bounded subsets in the quotient which are not contained in the closure of the canonical image of a bounded subset of E: choose a neighborhood of 0 bounded in the quotient space; then the strong dual of the quotient of E cannot be identified with a topological vector subspace of the strong dual of E). There is a larger class of reflexive quasi-normable spaces which satisfy Corollary 2:

PROPOSITION 2   *Let E be an LCTVS. The following conditions are equivalent:*

1) *Every continuous mapping from E into a Banach space F is weakly compact (resp. compact).*

2) *For every disked neighborhood U of 0 in E there exists a disked neighborhood $V \subset U$ such that the natural mapping from $\hat{E}_V$ into $\hat{E}_U$ is weakly compact (resp. compact).*

3) *For every equicontinuous disk A in E' there exists an equicontinuous*

disk $B \supset A$ *such that* A *is weakly relatively compact (resp. relatively compact) in the normed space* E'_B.

If E satisfies these properties and is furthermore quasi-complete then it is reflexive (resp. of type $(\mathscr{M})$).

Notice that if U is a disked neighborhood of 0 in E, $\hat{E}_U$ stands for the Banach space associated with the semi-norm gauge of U. The equivalence of 1) and 2) is trivial (1) implies 2) as we can see setting $F = E_U$; 2) implies 1) considering the inverse image U of the unit ball of F by u). On the other hand, 3) means that the identity mapping from E'_A into E'_B is compact; supposing A and B weakly closed (which changes nothing) and calling U and V the polars of A and B, the identity mapping $E'_A \rightarrow E'_B$ is the transpose of the canonical mapping $\hat{E}_V \rightarrow \hat{E}_U$, therefore it is weakly compact (resp. compact) if and only if the last mapping is. This establishes the equivalence of conditions 2) and 3). Finally, if E satisfies these conditions and is quasi-complete we shall show that it is reflexive (resp. a Montel space), i.e. that the identity mapping $E \rightarrow E$ transforms bounded subsets into weakly relatively compact (resp. relatively compact) subsets. We know (Chapter 2, Section 18, Theorems 12 and 13) that this is equivalent to saying that the transpose, i.e. the identity mapping from E' onto E' transforms equicontinuous subsets into relatively compact subsets for $\sigma(E', E'')$ (resp. relatively compact for the strong topology) which follows from Condition 3.

It is trivial that a quotient space of a space E satisfying the conditions of Proposition 2 also satisfies these conditions.

Using Corollary 2 of Theorem 2 we obtain:

COROLLARY 1   *Let* E *be a reflexive quasi-normable space (resp. of type* $(\mathscr{M})$). *Then every quasi-complete Hausdorff quotient of* E *is also quasi-normable and reflexive (resp. of type* $(\mathscr{M})$).

By condition 2) or condition 3) of Proposition 2, we see that the property considered in this proposition is closed under formation of arbitrary vector subspaces. In particular:

COROLLARY 2   *Let* E *be a quasi-normable and reflexive LCTVS (resp. of type* $(\mathscr{M})$). *Then every vector subspace of* E *satisfies the conditions of Proposition 2.*

We remark that this subspace may not be quasi-normable. We find for example that every vector subspace of the topological vector product of a family of *reflexive* Banach spaces satisfies the conditions of

Proposition 2 (those relative to weak compactness, i.e. which are not in parenthesis). This is the case for the spaces $(\mathscr{D}_{L_p})$ of L. Schwartz for $1 < p < \infty$ which are in fact by definition isomorphic to topological vector subspaces of the product of a sequence of L^p spaces (which are reflexive Banach spaces). In fact, it is not hard to see that the $(\mathscr{D}_{L_p})$ are even quasi-normable, although not in the general case, e.g. its vector subspaces which nevertheless satisfy the conditions of Proposition 2.

4 Schwartz spaces

A *Schwartz space* is a Hausdorff LCTVS which satisfies the second series of equivalent conditions of Proposition 2. We can for example put Condition 2 in the form:

DEFINITION 2   *Let E be a Hausdorff LCTVS; E is a Schwartz space abbreviated (S) space if for every disked neighborhood U of 0, there exists another one, V, which is precompact for the semi-normed topology defined by U.*

As well as the equivalent conditions of Proposition 2 we have another characteristic of (S) spaces:

PROPOSITION 3   *Let E be a Hausdorff LCTVS. E is an (S) space if and only if it is quasi-normable and if every bounded subset is precompact.*

The condition is sufficient by Corollary 2 of Theorem 2. Suppose, conversely, that E is an (S) space; let A be a bounded subset of E. In order to show that A is precompact it suffices to show that it is precompact for every continuous semi-norm corresponding to a disked neighborhood U of 0; this is clear by Definition 2. Finally, let A be weakly a closed equicontinuous disk in E'; from Condition 3 of Proposition 2 there exists a weakly closed equicontinuous disk B such that A is compact in E'_B. Therefore, on A the topology induced by E'_B is identical to the topology induced by E' strong which is Hausdorff and coarser. It follows that E is quasi-normable.

The interest of (S) spaces is due mainly to their properties of closure and because in the case of $(\mathscr{F})$ spaces they allow the completion of strong duality theory as we shall now show.

PROPOSITION 4   *(S) spaces are closed under the formation of subspaces, Hausdorff quotient spaces, topological vector products and topological direct sums of sequences of such spaces.*

We have already pointed out that the conditions of Proposition 2 are closed under formation of vector subspaces and quotients. For the closure of the product use Condition 2 of Proposition 2 (a continuous linear mapping from the product into a Banach space is equivalent to a continuous linear mapping from a partial finite product into the same space). For the topological direct sum of a *sequence* of spaces, see Exercise 1. For our purpose here ((S) spaces) we can say that a topological direct sum of quasi-normable spaces is quasi-normable, and if the bounded subsets of the factors are precompact, so are the bounded subsets in the direct sum (which are in fact contained in the sum of a finite number of bounded subsets of the factors), the result follows by Proposition 3.

COROLLARY 1   *A Hausdorff LCTVS which is the inductive limit of a sequence of (S) spaces is an (S) space.*

COROLLARY 2   *A Hausdorff LCTVS whose topology is the coarsest for which linear mappings u_i from E into (S) spaces E_i are continuous, is an (S) space.*

In the first case we have the quotient of a topological direct sum of a sequence of (S) spaces, in the second case we have a subspace of the product of a family of (S) spaces. Notice that if E is an $(\mathscr{F})$ and $(\mathscr{M})$ space which is not an (S) space (we have said in Section 3 that some existed), E is the strong dual of its strong dual E', which is of type (S) since it is quasi-normable (as a quasi-barrelled $(\mathscr{DF})$ space) and of type $(\mathscr{M})$ (as the dual of a quasi-barrelled $(\mathscr{M})$ space). Then the strong dual of an (S) space can fail to be of type (S).

THEOREM 3   *Let E be an $(\mathscr{F})$ and (S) space, F a closed vector subspace. Then F and E/F are $(\mathscr{F})$ and (S) spaces, the strong dual of F (resp. of E/F) can be identified with the quotient E'/F^0 of the strong dual E' of E (resp. with the topological vector subspace F^0 of the strong dual E' of E).*

We know already (Proposition 4) that F and E/F are of type (S) and also $(\mathscr{F})$ spaces. In particular, F will be reflexive whence we obtain the result on the strong dual (Chapter 2, Section 15, Proposition 21, Corollary 2). For the assertion relative to the dual of E/F it suffices to show that every bounded subset of E/F is contained in the closure of the canonical image of a bounded subset of E. Now the bounded subsets of E/F being relatively compact it suffices to apply Part 2, Section 1, Proposition 1.

COROLLARY   *Let E be an $(\mathscr{F})$ and (S) space. Then the closed vector sub-*

N

spaces of its strong dual and the quotient spaces of the dual by the subspaces are $(\mathscr{D}\mathscr{F})$, (S), *bornological and complete spaces.*

They are strong duals of $(\mathscr{F})$ and (S) spaces therefore they are $(\mathscr{D}\mathscr{F})$ spaces, complete and bornological (Part 3, Section 5, Theorem 6, Corollary 3). Since they are also of type $(\mathscr{M})$, they are of type (S) as we have pointed out above for the barrelled $(\mathscr{D}\mathscr{F})$ spaces.

PROPOSITION 5   *Let U be an open subset of $\mathbf{R}^n$, K a compact cube of $\mathbf{R}^n$. Then the spaces $\mathscr{E}(U)$ and $\mathscr{E}(K)$ of indefinitely differentiable functions on U and K (see Chapter 1, Section 10) are (S) spaces.*

In fact, the topology of $\mathscr{E}(K)$ for example is the coarsest for which the identity mappings from $\mathscr{E}(K)$ into the spaces $\mathscr{E}^{(m)}(K)$ are continuous. Now we have seen (Chapter 1, Section 10) that the identity mapping from $\mathscr{E}^{(m+1)}(K)$ into $\mathscr{E}^{(m)}(K)$ is compact, whence the condition of Definition 2 is verified. We proceed analogously for $\mathscr{E}(U)$.

COROLLARY 1   *The space $\mathscr{D}(U)$ of indefinitely differentiable functions with compact support in U is an (S) space.*

For every compact $K \subset U$, the space $\mathscr{D}_K(U)$ of $f \in \mathscr{E}(U)$ whose support is in K is a topological vector subspace of $\mathscr{E}(U)$, thus an (S) space. $\mathscr{D}(U)$ is the inductive limit of a sequence of such spaces $\mathscr{D}_K(U)$ of type (S), thus itself an (S) space.

COROLLARY 2   *Let U be an open set of $\mathbf{C}^n$; the space $\mathscr{H}(U)$ of holomorphic functions on U with the topology of compact convergence is an (S) space. So is the space $\mathscr{H}(K)$ of functions holomorphic in the neighborhood of a compact K of $\mathbf{C}^n$ (Part 1, Section 2, Example f).*

We know that $\mathscr{H}(U)$ is a topological vector subspace of $\mathscr{E}(U)$ (use the closed graph theorem and the fact that uniform limits on every compact set of holomorphic functions are holomorphic, which a fortiori means that $\mathscr{H}(U)$ is a *closed* subspace of $\mathscr{E}(U)$ which is of type (S), then so is $\mathscr{H}(U)$. Finally, $\mathscr{H}(K)$ is the inductive limit of a sequence of $\mathscr{H}(U)$ which are (S) spaces therefore an (S) space.

EXERCISE 1   Let E be an $(\mathscr{F})$ space and let (A_i) be a sequence of weakly compact (resp. compact) disks in E. Show that we can find a sequence (λ_i) of scalars > 0 such that the closed disked hull of the $\cup \lambda_i A_i$ is weakly compact (resp. compact). Conclude that if F is an LCTVS inductive limit of a sequence of LCTVS F_i by linear mappings u_i whose images generate F, and u a linear mapping from F into E (of type $(\mathscr{F})$), then u is weakly compact (resp. compact, resp. bounded) if and only if the mappings $u \circ u_i$ are weakly compact (resp. compact, resp. bounded). Conclude that the conditions of Proposition 2 are closed under passage to the inductive limit of a sequence of spaces.

EXERCISE 2   Consider the situation of Part 3, Section 3, Proposition 5, Corollary 2. Show that if the mappings v_i are weakly compact (resp. compact) then the strong dual of E is an ($\mathscr{F}$) space satisfying the first (resp. the second) series of conditions of Proposition 2 of the present section. Then, E is the strong dual of a reflexive ($\mathscr{F}$) space (resp. of an ($\mathscr{F}$) and (S) space).

CHAPTER 5

Compactness in locally convex topological vector spaces (LCTVS)

THE MOST ELEMENTARY properties of compactness (and the most important ones) referring to the duality technique have been seen in Chapter 2, Section 18. Here we present more refined properties. Parts 1 and 2 are important for certain applications (beyond the theory of TVS itself). They are independent from each other and from Parts 3 and 4.

PART 1 THE KREIN–MILMAN THEOREM

1 Extreme points

The vector spaces considered here will be *real*. The segment with end points a and b is the set of points $\lambda a + (1 - \lambda)b$ with $0 \leqslant \lambda \leqslant 1$; for convenience we call the segment without its end points the interior of the segment.

DEFINITION *Let A be a subset of a vector space. A linear sub-variety V of E is a support variety if $V \cap A \neq \phi$ and if every open segment contained in A whose interior meets V is contained in V. We call a linear support variety of dimension 0 an extreme point of A (i.e. a point of A which is an end point of every segment in A which contains it).*

PROPOSITION 1 *An intersection of support varieties of A is a support variety if and only if it meets A*

The proof is trivial.

COROLLARY *If A is a compact subset of a TVS E, the set of closed support varieties of A is inductive for $\supset$.*

For reasons of compactness the intersection of a totally ordered family of closed support varieties meets A, therefore it is a closed support variety which is the l.u.b.

PROPOSITION 2 *Let A be a subset of E, a vector space, let x' be a non-zero* *linear form on E, V a hyperplane of equation*

$$\langle x, x' \rangle = a$$

V is a support variety of A if it meets A and "keeps it on one side" i.e. a a minimum or a maximum of the form x' on A. The condition is also necessary if A is convex.

The proof is trivial.

COROLLARY *If A is a compact subset of a TVS then for every closed hyperplane V in E, there exists a parallel support hyperplane of A.*

V will have an equation $\langle x, x' \rangle = 0$, where x' is a *continuous* linear form on *E* which then admits a maximum *a* on *A*. The hyperplane of equation $\langle x, x' \rangle = a$ satisfies the condition.

PROPOSITION 3 *Let A be a subset of E, V a support variety of A, W a variety contained in V. Then W is a support variety of A if and only if it is a support variety of V ∩ A.*

The necessity follows from the trivial fact that every support variety *W* of *A* is also a support variety of every subset of *A* that meets *W*; the sufficiency from: a segment in *A* whose interior meets *W* is contained in *V*, therefore in *V* ∩ *A* therefore in *W* which is a support variety of *V* ∩ *A*.

From Propositions 2 and 3 follows the

COROLLARY *Let A be a compact subset of a Hausdorff LCTVS E. Then the extreme points of A are the minimal closed support varieties.*

Let *V* be a closed support variety not reduced to a point. We shall show that there exists a closed support variety strictly smaller. We can suppose 0 ∈ *V* (by translation if necessary), then *V* ∩ *A* is a non-empty compact subset of the non-zero Hausdorff LCTVS *V*. In *V* there exists a closed hyperplane (Hahn–Banach) therefore a parallel support hyperplane of *A* ∩ *V* (Proposition 2, Corollary) which is also a variety of support of *A* (Proposition 3).

From the preceding corollary and from the Corollary of Proposition 1 which allows Zorn's theorem to be applied) we have:

PROPOSITION 4 *Let A be a compact subset of Hausdorff LCTVS E. Every closed support variety of A contains at least one extreme point.*

Since *E* is a support variety of *A*, *A* has at least one extreme point. Better still:

THEOREM 1 (KREIN–MILMAN) *Let A be a compact subset of a Hausdorff*

LCTVS E. *Then the set of extreme points of A has the same closed convex hull as A.*

In the case where A is convex (apparently the most interesting) this statement becomes

COROLLARY   *A compact convex subset of a Hausdorff* LCTVS *is identical to the closed convex hull of the set of its extreme points.*

Proof of Theorem 1   By Chapter 2, Theorem 3, Corollary 1, it suffices to show that if $x' \in E'$ and $a \in \mathbf{R}$ are such that

$$\langle x, x' \rangle \leqslant a$$

for every extreme point of A, we have the same inequality for every $x \in A$. If not, the maximum of x' on the compact A would be $b > a$; now the hyperplane $\langle x, x' \rangle = b$ would be a support hyperplane of A (Proposition 2), therefore would contain an extreme point (Proposition 4), a contradiction, since at that point x' has the value b not $\leqslant a$.

2 Extreme generators

For the elementary study of cones, see Chapter 2, Section 3. Here we only consider cones containing 0. Let V be a support variety of the cone C; if it contains a point x of C, it contains the generator of x (since it contains a segment carried by the generator whose interior contains x). Therefore in all cases, since for a support variety V, $V \cap C$ is non-empty, a support variety of the cone C contains 0, i.e. it is homogeneous. We are interested in support varieties which contain at least one generator, i.e. whose intersection with C is not reduced to zero.

DEFINITION 2   *Let C be a cone in the vector space E. A generator of C is an extreme generator if the line it generates is a support variety.*

If x is a point of C, its generator is an extreme generator if and only if every segment in C whose interior contains x is contained in the generator of x, i.e., if $y, z \in A$, $0 < \lambda < 1$ and $x = \lambda y + (1 - \lambda)z$ implies that y and z are proportional to x. (The necessity is clear by definition, the sufficiency is also clear if we notice that if the condition is verified it is also verified for the points of C proportional to x.)

Let C be a cone in the vector space E, H a non-homogeneous hyperplane in E such that C is the cone (containing the origin) generated by $A = C \cap H$. Then the vector subspaces W of E such that $W \cap C$ is not reduced to zero are in bijective correspondence with the linear sub-varieties V of H meeting A, assigning $V = W \cap H$ to W and to V the vector space W generated by V.

PROPOSITION 5 *In this correspondence, the support varieties W of C such that $W \cap C \neq 0$, correspond to the support varieties V of $A = H \cap C$.*

Suppose W is a support variety of C; consider a segment in A whose interior meets V, it is then contained in C and meets W, therefore it is contained in W, thence also in $V = W \cap H$; thus V is a support variety of A. Suppose that V is a support variety of A; consider a segment in C whose interior contains an $x \in W$, $x \neq 0$ (because of the way C is generated). Either this segment is on the line generated by x (a fortiori contained in W) or it is projected (centrally) on H on a segment contained in A whose interior contains the projection of x on H and thus meets $V = W \cap H$; this last segment is then contained in V, whence the initial segment is in W, which proves that W is a support variety of C.

COROLLARY *Let C be the convex cone containing 0, generated by a subset A of a non-homogeneous hyperplane H. The extreme generators of C are in bijective correspondence with the extreme points of A (to an extreme point of A corresponds the generator of C generated by this point).*

We have seen in Chapter 2, Part 3, that the existence of a convex cone C containing the origin in a vector space E, is equivalent to the existence of a pre order on E compatible with the vector structure (the cone being the set of elements $\geqslant 0$ in E).

PROPOSITION 6 *Let E be a real pre-ordered space, C the cone of its positive elements. Let $x \in E$, $x > 0$. The generator of x is an extreme generator of C if and only if every positive element of E bounded above by x is proportional to x.*

If the generator of x is not an extreme generator we have

$$x = \lambda a + (1 - \lambda)b$$

with $a, b \in C$, $0 < \lambda < 1$, a and b not proportional to x (see remark after Definition 2). Then λa is a positive element of E, bounded above by x and not proportional to x. Suppose there exists such an element y of E, $0 \leqslant y \leqslant x$, y not proportional to x; then we have

$$x = \tfrac{1}{2}(2y + 2(x - y)),$$

and we are back to the initial conditions with $a = 2y$, $b = 2(x - y)$ $\lambda = \tfrac{1}{2}$, so that the generator of x is not extreme. It is mainly in the form of Proposition 6 that the extreme generaters of convex cones appear in applications.

Up to this point we have not dealt in this Section with a topology on E, which we will now do:

THEOREM 2 *Let E be a Hausdorff LCTVS, A a compact convex subset of E not meeting 0, C the cone containing the origin, generated by A. Then C is closed and identical to the closed convex hull of the union of its extreme generators.*

We can find a closed hyperplane H_0 not meeting A and consequently keeping A strictly to one side. Let H be a hyperplane parallel to H_0 of equation $\langle x, x' \rangle = 1$, of the same side of H_0 as A. For every $x \in \complement H_0$ let

$$p(x) = \frac{1}{\langle x, x' \rangle} x$$

be its central projection on H. The cone containing 0 generated by A is identical to the cone containing 0 generated by $K = p(A)$ and except for the origin it is the set of points x of the open half-space U defined by $\langle x, x' \rangle > 0$ such that $p(x) \in K$. Now as p is a continuous mapping in $\complement H_0$, K will be compact since it is the image of the compact A. A fortiori K is closed, then its inverse image in U by p which is C minus the origin is relatively closed in U. Then every point in the closure of $C \cap U$ not contained in $C \cap U$ is not contained in U, thence not in H_0, and we shall show it is zero. This follows from the immediate fact that we can strictly separate every non-zero point x of H_0 from A by some closed hyperplane whence x is not in the closure of the cone generated by A. We have thus proved that C is closed. By Theorem 1 the compact K is the closed convex hull of the set of its extreme points, therefore contained (taking into consideration the corollary of Proposition 5) in the closed convex hull of the union of extreme generators of C; this is equally true of C, which ends the proof.

EXERCISE 1

1) Let M be a locally compact space. Let $\mathscr{K}(M)$ be the ordered vector space of real continuous functions with compact support in M, $\mathscr{M}^+(M)$ the cone of positive linear forms on $\mathscr{K}(M)$. Show that the extreme generators are generated by the linear forms $\varepsilon_s (s \in M)$ defined by $\langle f, \varepsilon_s \rangle = f(s)$.

2) Let $\mathscr{M}^1(M)$ be the dual of the Banach space $C_0(M)$. Let A be the set of elements of the unit ball of $\mathscr{M}^1(M)$ which are positive linear forms on the ordered vector space $C_0(M)$. Show that the extreme points of A are 0 and the ε_s.

3) If M is compact, show that the set K of elements of A of norm 1 is identical to the set of positive linear forms μ on $C(K)$ such that $\mu(1) = 1$. Conclude that K is weakly compact. Show that its extreme points are the ε_x and that the cone containing 0 generated by K is identical to $\mathcal{M}^+(M)$.

4) Apply the Krein–Milman theorem to the situations in 2) and 3) and notice also that here we obtain the same results more rapidly by direct application of the bipolar theorem, compare with Chapter 2, Section 9, Exercise).

5) Consider furthermore $C_0(M)$ (M locally compact), the scalars being either real or complex. Show that the extreme elements of the unit ball of the dual $\mathcal{M}^1(M)$ are the $\lambda \varepsilon_s$ with $s \in M$, $|\lambda| = 1$.

EXERCISE 2   (The scalars are real or complex)

1) Let M be a topological space. Show that the extreme points of the unit ball of $C^\infty(M)$ are the f such that $|f(s)| = 1$ for every $s \in M$. Prove the analogous assertion for a space L^∞ constructed on a measure μ.

2) Show that in this last case, the unit ball of L^∞ is the closed convex hull of the set of its extreme points. Show that this is also true for $C(M)$ where M is compact and totally discontinuous. (The Krein–Milman theorem does not work in this case! We must consider a space $C(M)$ where M is a compact finite set. We can remark that, as is well known, L^∞ is isomorphic to a $C(M)$ constructed on a certain totally discontinuous compact space M; therefore, in fact, the assertion on L^∞ can be considered as a particular case of that relative to $C(M)$.)

3) Let M be a *connected* topological space; show that the unit ball of $C^\infty(M)$ when we choose real scalars has only two extreme points and is not in general the closed convex hull of the set of its extreme points.

EXERCISE 3   Let u be a linear mapping from one vector space E into another, F. Let A be a subset of E, let $B = u(A)$. Show that the inverse image by u of a support variety of B is a support variety of A. Conclude that if E, F are Hausdorff LCTVS, A compact, then every extreme point of B is the image by u of an extreme point of A; and if A is the cone containing the origin generated by a compact convex set not containing 0, then every extreme generator of B is the image by u of an extreme generator of A.

EXERCISE 4   Let E be a finite dimensional vector space over $\mathbf{R}$, K a compact convex subset of E. Show that K is the convex hull of its extreme points (in order to show that every $x \in K$ is in this hull, take x in the interior of K then reason by induction on the dimension).

EXERCISE 5   Let E be a Hausdorff LCTVS over $\mathbf{R}$, A a subset of E.

a) Show that the following conditions on an $x \in E$ are equivalent:

1. For every continuous linear mapping u from E into a finite dimensional vector space, $u(x)$ belongs to the convex hull of $u(A)$.

2. For every closed subvariety H of E, of finite codimension $n > 0$, containing x, there exists a subset A having at most $n + 1$ elements whose convex hull meets H (use Chapter 2, Section 5, Exercise 2). Let $\tilde{A}$ be the set of $x \in E$ that satisfy these properties. $\tilde{A}$ is convex and we have $A \subset \tilde{A} = \tilde{\tilde{A}}$ and $\tilde{A}$ is contained in the closed convex hull of A. If E is finite dimensional, A is identical to the convex hull of A.

b) Let (x_i) be a family of elements of A; (λ_i) a family of positive numbers such that $\Sigma \lambda_i = 1$, and the family of $(\lambda_i x_i)$ in E is summable. Show that its sum x is in A. (Reduce to the case where E is finite dimensional and is generated affinely by the $\lambda_i x_i$). More generally, if M is a locally compact space with a positive measure μ of total mass 1, $t \mapsto f(t)$ a scalarly summable mapping from M into E such that

$$x = \int f(t) \, d\mu(t) \in E,$$

if we suppose that $f(M) \subset A$ then $x \in \tilde{A}$. (Reduce to the case where E is finite dimensional, then to the preceding case, writing M as $N \cup (\underset{i}{\cup} M_i)$ where N is negligible, and where (M_i) is a sequence of pairwise disjoint compact s on which f has a continuous restriction—using Chapter 2, Section 5, Exercise 2.)

c) If A is compact, $\tilde{A}$ is identical to the closed convex hull of A (use Chapter 2, Section 5, Exercise 2).

d) If K is compact convex subset of E, A the set of its extreme points we have $K = \tilde{A}$ (use Exercises 3 and 4).

e) If A is a subset of E which is the union of closed half-lines of origin 0, then Condition 1 of a) can be replaced by the following: for every closed variety H of finite codimension n containing x, there exists a finite subset of A having *at most n elements* whose convex hull meets H (or also: whose sum equals x). Furthermore, the sum of *every* summable family extracted from A is in A, then generalize as in b) for the integrals of weakly summable functions. If A is the union of extreme generators of a cone C generated by a compact convex set K not containing the origin, we have $\tilde{A} = C$.

PART 2   THEORY OF COMPACT OPERATORS

1 Generalities

In this Part we consider *Hausdorff* LCTVS only.

Recall (Chapter 2, Section 18, Definition 14) that a linear mapping from a LCTVS E into another one, F, is compact (resp. precompact) if it transforms some neighborhood V of 0 into a relatively compact (resp. precompact) subset; a fortiori it is continuous. When E is normed we can choose V to be the unit ball in the definition; if F is quasi-complete the notions of compact mapping and precompact mapping coincide. Taking into consideration Chapter 0, Section 4, Proposition 6' it follows that if E is normable, F quasi-complete, the space of compact linear mappings from E into F is a closed vector subspace of $L_b(E, F)$. Recall also that the transpose of a compact linear mapping of a Banach space into another is compact (Chapter 2, Section 18, Theorem 12, corollary 3). We point out that this is more generally true for the transpose of a compact linear mapping from an LCTVS into an ($\mathscr{F}$) space when we equip the duals with the strong topology (see Chapter 4, Part 2, Exercise 4).

We also point out.

PROPOSITION 1   *Let u be a continuous linear mapping from one LCTVS E into another, F. Let E'_c and F'_c be the duals of E and F, with the topology of uniform convergence on the compact disks. If u is compact, then its transpose u' is also compact.*

Let V be a disked neighborhood of the origin in E such that $K = u(V)$ is relatively compact, then u' maps K^0 into V^0 (Chapter 2, Proposition 25). Now K^0 is a neighborhood of 0 in F'_c and V^0 in an equicontinuous subset of E', therefore relatively compact for the weak topology, and also for compact convergence (and a fortiori for uniform convergence on the compact disks) by reasons of equicontinuity. This ends the proof; notice that the converse is clearly true if every compact disk in E'_c is equicontinuous, in particular, if the topology of E is $\tau(E, E')$.

2 General theorems for finite dimension

Here we follow almost word for word a recent note of L. Schwartz.

THEOREM 1   *Let u, v be two linear mappings from one LCTVS E into another, F. We suppose that u is an isomorphism from E onto a closed* *subspace of F and that v is compact. Then w = u + v is a homomorphism whose image is closed and whose kernel is finite dimensional.*

Let V be a disked neighborhood of 0 such that $v(V)$ is relatively compact, let N be the kernel of w, let $W = V \cap N$. Then we have

$$u(W) = -v(W) \subset v(V).$$

Thus $u(W)$ is relatively compact therefore precompact, whence W is precompact (u is an isomorphism). Thus N has a precompact neighborhood W of 0, therefore it is finite dimensional (Chapter 1, Section 13, Theorem 8). It is clear that in order to prove the other assertions of the theorem we can choose the restrictions of u, v, w to a topological supplement of N (it exists as N is finite dimensional, see Chapter 2, Section 5, Hahn–Banach theorem, Corollary 4), therefore we are back to the case where w is bijective. We must now show that w is an isomorphism onto a closed subspace of F; it is immediate that this means that if U is an ultrafilter in E such that $\lim_U wx$ exists in F, then U converges in E. Let p be the semi-norm gauge of V, let $a = \lim_U p(x)$ $(0 \leqslant a \leqslant +\infty)$. If a is finite then there exists $A \in U$ such that A is contained in a dilation $(a + 1)V$, then $v(A)$ is relatively compact, therefore $\lim_U vx$ exists and $\lim_U ux = z$ (since $ux = wx - vx$). But, E being closed, we have $z = \mu x_0$ $(x_0 \in E)$ and u being an isomorphism we have $\lim_U x = x_0$. We shall show that it is impossible to have $a = +\infty$. If not, we would have

$$\lim_U \frac{wx}{p(x)} = \lim_U w\left(\frac{x}{p(x)}\right) = 0,$$

or, by the preceding argument, $x/p(x) \to x_0$ and we would have

$$p(x_0) = \lim_U p\left(\frac{x}{p(x)}\right) = 1$$

and $w(x_0) = 0$, which contradicts the bijectivity of w.

THEOREM 2 *Let u, v be linear mappings from one LCTVS E into another, F. We suppose that u is a weak homomorphism from E onto F such that every compact disk of F is contained in the image by u of a compact disk of E and that v is compact. Then w = u + v is a weak homomorphism from E onto a closed subspace of finite codimension of E.*

Equip E, F' with the topology of uniform convergence on compact disks (their duals are then E and F), v' is compact (Proposition 1), $u'(F')$ is weakly closed since u is a weak homomorphism (Chapter 2, Section 16, Proposition 27) thence closed, finally u' is continuous (Chapter 2,

Proposition 28) and even an isomorphism; since every equicontinuous subset of the dual F of F' i.e. every compact disk of F is by hypothesis contained in the image by $(u')' = u$ of an equicontinuous subset of the dual of E', i.e. of a compact disk of E, so that it suffices to apply Chapter 1, Proposition 29, Corollary 1. Thus u' and v' satisfy the hypothesis of Theorem 1, hence $w' = u' + v'$ is a homomorphism from F' onto a closed subspace of E' having a finite dimensional kernel. Thus w is a weak homomorphism (since $w(F')$ is weakly closed) and $w(E)$ is closed (w' being a homomorphism therefore a weak homomorphism) and even identical to the orthogonal of the kernel of w', thence of finite codimension, which ends the proof.

COROLLARY 1   *Let u, v be continuous linear mappings from a space E of type ($\mathscr{F}$) into another, F. We suppose that u is onto, v compact. Then $w = u + v$ is a homomorphism from E onto a closed subspace of finite codimension of F.*

The mapping u is a homomorphism (theorem of homomorphisms, Banach theorem, Chapter 1, Section 14, Theorem 9), a fortiori a weak homomorphism, the condition on lifting of compacts in the statement of Theorem 2 is automatically satisfied (Chapter 4, Part 2, Section 1, Proposition 1). Theorem 2 can be applied and from the fact that w is a weak homomorphism it follows that it is even a homomorphism (Chapter 4, Part 2, Section 4, Theorem 3).

COROLLARY 2   *Let E, F, G be ($\mathscr{F}$) spaces, u a continuous linear mapping from E into G, v a compact linear mapping from F into G such that $u(E) + v(F) = G$. Then u is a homomorphism from E onto a closed subspace of finite codimension of G.*

Let $H = E \times F$; let $\tilde{u}$ and $\tilde{v}$ be linear mappings from H into G defined by u and v ($\tilde{u}(x, y) = ux$, $\tilde{v}(x, y) = vy$), let $\tilde{w} = \tilde{u} + \tilde{v}$ whence $\tilde{u} = \tilde{w} + (-\tilde{v})$. Then $\tilde{w}$ is onto (by $u(E) + v(F) = G$) and $-\tilde{v}$ is compact, then by virtue of Corollary 1 $\tilde{u}$ is a homomorphism from H onto a closed subspace of finite codimension of G. This is clearly also true of u.   Combining theorems 1 and 2 we obtain:

COROLLARY 3   *Let u, v be linear mappings from one LCTVS E into another, F. We suppose that u is an isomorphism from E onto F, v compact. Then $w = u + v$ is a homomorphism from E onto a closed subspace of finite codimension of F, whose kernel is finite dimensional.*

In Section 4, we shall see that the dimension of the kernel of w is equal to the codimension of the image (we suppose $E = F$, u is the identity mapping from E onto F, a hypothesis which does not restrict the generality).

3 Generalities on the spectrum of an operator

We begin with some generalities of an algebraic nature (valid for vector spaces over an arbitrary field).

PROPOSITION 2 *Let v be an endomorphism of a vector space E. Let E_n be the kernel, F_n the image of v^n (for n an integer $\geqslant 0$), in particular*

$$E_0 = 0, \quad F_0 = E.$$

1) *The sequence (E_n) is either strictly increasing or strictly increasing up to a finite rank $n = v$, and constant from there on.*

2) *The sequence (F_n) is either strictly decreasing or strictly decreasing up to a finite rank μ, and constant from there on.*

3) *If the two sequences (E_n) and (F_n) end up being stationary then $v = \mu$ (with the notations of 1 and 2, i.e. they remain stationary beginning at the same rank. Now set $E_\infty = E_\nu$, $F_\infty = F_\nu$, then E is direct sum of E_∞ and F_∞ and v induces a nilpotent endomorphism in the first factor, an automorphism in the second factor.*

Proof We have

$$(v^p)^{-1}(E_q) = E_{p+q}$$
$$v^p(F_q) = F_{p+q}$$

on the other hand, since v^n permutes with v, its kernel E_n and image F_n are stable under v. Therefore

$$E_{n+1} = v^{-1}(E_n) \supset E_n$$
$$F_{n+1} = v(F_n) \supset F_n$$

whence (E_n) is increasing, (F_n) decreasing. If we have $E_n = E_{n+1}$ we have

$$(v^p)^{-1}(E_n) = (v^p)^{-1}(E_{n+1}),$$

$E_{n+p} = E_{n+p+1}$, i.e. the sequence (E_n) is stationary starting at n, and we see in the same way that if $F_n = F_{n+1}$, the sequence (F_n) is stationary starting at n, which ends the proof of 1 and 2. In the notations of 3, we would have shown $\mu \leqslant v$ if we prove that $E_n = E_{n+1}$ and $F_n \neq F_{n+1}$ implies $F_{n+1} \neq F_{n+2}$ (whence $F_{n+1} \neq F_{n+2}$ since we have $E_{n+1} = E_{n+2}$, so that (F_m) would be a strictly increasing sequence). If in fact we had $F_{n+1} = F_{n+2}$, letting $v^n x$ be an element of F_n, then

$$v^{n+1}x \in F_{n+1} = F_{n+2}$$

is of the form $v^{n+2}y$, then

$$v^{n+1}(x - Vy) = 0$$

and (since $F_n = F_{n+1}$)

$$v^n(x - vy) = 0$$

i.e.

$$v^n x = v^{n+1}y \in F_{n+1},$$

we would then have $F_n = F_{n+1}$, contrary to the hypothesis. We prove analogously that $F_n = F_{n+1}$ and $E_n \neq E_{n+1}$ implies $E_{n+1} \neq E_{n+2}$, whence $v \leqslant \mu$, and $\mu = v$. We have $E_\infty \cap F_\infty = 0$; since if x belongs to this intersection we have $x = v^v y$, $v^v x = 0$, whence $v^{2v}y = 0$ and $v^v y = 0$ (since $E_v = E_{2v}$), i.e. $x = 0$. Let v be the operator in E/E_∞ deduced from v by passing to the quotient, $\tilde{v}$ is clearly bijective (since $E_v = E_{v+1}$), and we shall show it is *onto*. In fact, from 3, applied to $\tilde{v}$ for which the integer corresponding to v in this statement is 0, the sequence of $v^n(E/E^\infty)$ would be strictly decreasing, whence the sequence of their inverse images

$$v^n(E) + E_\infty$$

would be strictly decreasing, and the sequence of the $V^n(E)$ could not remain stationary, contrary to the hypothesis. Thus $\tilde{v}$ is an automorphism, so is $\tilde{v}^v$, in particular

$$\tilde{v}^v(E/E_\infty) = E/E_\infty,$$

which can also be written

$$v^v(E) + E_\infty = E,$$

i.e. $F_\infty + E_\infty = F$. This proves that E is direct sum of F_∞ and E_∞. To say that $\tilde{v}$ is an automorphism is to say also that v induces an automorphism onto F_∞. On the other hand, since v^v is zero on E_∞, the restriction of v to E_∞ is nilpotent, the conclusion follows:

COROLLARY *Suppose we have the conditions of 3) and that E_∞ is finite dimensional. Then the dimension of the kernel of v is equal to the co-dimension of its image.*

It suffices to verify this separately for the restrictions of v to E_∞ and F_∞. Now in the finite dimensional space E_∞ this is well known, also, v induces an automorphism of F_∞, consequently the dimension of its kernel and the codimension of its image are both zero.

Recall that the *spectrum* of an element u of an algebra A with unity 1 is the set of scalars λ such that $\lambda 1 - u$ cannot be inverted. In particular, if u is a linear operator in a vector space E, when we mention its spectrum, we understand the spectrum of u in the algebra of endomorphisms of E, or of all *continuous* endomorphisms of E when E is a TVS; if E is an $(\mathscr{F})$ space, the spectrum of u in either of the two algebras is the same since we know that if a continuous operator $v = 1 - u$ can be inverted in the set of linear endomorphisms, i.e. if it is bijective, then its inverse is continuous (Theorem of homomorphisms, Chapter 1, Section 14, Theorem 9). Let u be an endomorphism of a vector space, a scalar λ is an *eigen value* of u if there exists a non-zero x in E such that $ux = \lambda x$; since this can be written $(\lambda 1 - u)x = 0$, we see that this is equivalent to saying that $\lambda 1 - u$ is not bijective, a fortiori cannot be inverted, thus an eigenvalue is a spectral value (the converse being in general false if E is infinite dimensional). The $x \in E$ such that $ux = \lambda x$ are called the *eigenvectors* associated with the eigenvalue λ. The space of eigenvectors relative to λ is then also the kernel of

$$\lambda 1 - u = v_\lambda.$$

Generally, for every scalar λ, set $v_\lambda = \lambda 1 - u$, then

$$E_{\lambda,n} = v_\lambda^{-n}(0)$$
$$F_{\lambda,n} = v_\lambda^n(E)$$

$$E_\lambda = \bigcup_n E_{\lambda,n} \qquad F_\lambda = \bigcup_n F_{\lambda,n}$$

The sequence $(E_{\lambda,n})$ is increasing, $(F_{\lambda,n})$ decreasing; for $\lambda = 0$ we find the sequences of spaces of Proposition 2 for the operator u. E_λ is also called the *spectral subspace* of E relative to λ, its dimension is called the *spectral multiplicity* of λ.

PROPOSITION 3   *If λ and λ' are distincts scalars then $E_\lambda \subset F_{\lambda'}$.*

Let $x \in E_\lambda$, i.e. $(\lambda 1 - u)^n x = 0$ for some n, we must show that we can write for any m,

$$x = (\lambda' 1 - u)^m y.$$

Now, let G be the vector space stable under u (or $\lambda 1 - u$, which amounts to the same) generated by x, which has *finite* dimension $\leqslant n$; the restriction of u to G having as only eigenvalue λ, $\lambda' 1 - u$ has a restriction to G which is invertible; this also holds for $(\lambda' 1 - u)^m$, consequently

$$x \in (\lambda' 1 - u)^m G = G,$$

which ends the proof.

Recall that if E is finite dimensional and the field of scalars algebraically closed (for example the field of complex numbers), then E is the direct sum of the spectral subspaces E_λ corresponding to the eigenvalues

of u which are the roots of the characteristic polynomial
$$P_u(\lambda) = \det(\lambda 1 - u),$$
of degree $n = \dim E$. The order of multiplicity of a root equals its spectral multiplicity.

PROPOSITION 4   *Let E be a Banach space. The set of invertible elements of $L(E)$ is open and contains the set of $1 - u$ with $\|u\| < 1$.*

Proof   If $\|u\| < 1$, the series $\sum_0^\infty u^n$ is absolutely convergent and we verify trivially that its sum is the inverse of $1 - u$. Let v be an invertible element of $L(E)$, then for $w \in L(E)$,
$$v + w = v(1 - (-v^{-1}w))$$
is invertible if $\|v^{-1}w\| < 1$, a fortiori if $\|v^{-1}\|\,\|w\| < 1$, therefore in every case for w sufficiently small, which ends the proof.

COROLLARY   *For every $u \in L(E)$, the spectrum $\sigma(u)$ of u is a compact subset of the field of scalars; $\lambda \in \sigma(u)$ implies $|\lambda| \leqslant \|u\|$.*

By definition, $\sigma(u)$ is the inverse image by the mapping $\lambda \mapsto \lambda 1 - u$ of the set of non-invertible elements of $L(E)$, and since this last one is closed by Proposition 3, so is $\sigma(u)$. Furthermore if $\lambda > \|u\|$, then by Proposition 3, $1 - (1/\lambda)u$ is invertible, and so is $\lambda 1 - u$, and λ does not belong to $\sigma(u)$. In other words, $\lambda \in \sigma(u)$ implies $\lambda \leqslant \|u\|$. Thus the spectrum of u is bounded and therefore compact since it is closed. Notice that Proposition 3 and its corollary remain valid for a complete normed algebra with unity.

If u is a continuous operator in a non-normable LCTVS its spectrum is not in general either closed or bounded (see Exercise 1). We can however in certain cases reduce the problem to the case of a Banach space by

PROPOSITION 5   *Let E be an LCTVS, F a vector subspace of E with a locally convex topology finer than the topology induced by E. Let u be a continuous mapping from E into F and call ϕ the inclusion mapping from F into E. Then with the exception of the scalar 0, the spectrum of ϕu in $L(E)$ is identical to the spectrum of $u\phi$ in $L(F)$. If λ is a non-zero element of the spectrum, then the eigensubspace and the corresponding spectral subspace is the same for $u\phi$ and ϕu.*

We must first show that if λ is a non-zero scalar, to say that $\lambda 1 - \phi u$ is invertible in $L(E)$ is to say that $\lambda 1 - u\phi$ is invertible in $L(F)$. Dividing by the scalar λ and replacing in the notations $-(1/\lambda)u$ by u, it suffices to prove this for $1 + \phi u$ and $1 + u\phi$. Now suppose that $1 + \phi u$

o

has an inverse in $L(E)$ which we can write $1 + v$, $v \in L(E)$, then

(1) $$\phi u + v + \phi u v = \phi u + v + v \phi u = 0.$$

From the first relationship in (1) we have

$$v = \phi w \text{ where } w = -u(1 + v) \in L(E, F)$$

and replacing v in (1) we find

$$\phi(u + w + u\phi w) = \phi(u + w + w\phi u) = 0$$

and, suppressing the factor ϕ to the left (which is possible since ϕ is injective) and multiplying the relations obtained to the right by ϕ:

$$u\phi + w\phi + (u\phi)(w\phi) = u\phi + w\phi + (w\phi)(u\phi) = 0$$

which means precisely that $1 + w\phi \in L(F)$ is an inverse of $u\phi$ in $L(F)$. Similarly we can show that if $u\phi$ is invertible so is u. Let λ be a non-zero spectral value of $u\phi$, ϕu, then it is obvious that the eigensubspace corresponding to u in F is the intersection of F with the eigensubspace of E corresponding to ϕu; however, the latter is already contained in F, since $\phi u x = \lambda x$ implies $x = (1/\lambda)\phi u x \in F$, the identity of eigensubspaces then follows. From this we conclude more generally that for every n, the kernels of $(\lambda 1 - \phi u)^n$ in E and of $(\lambda 1 - u\phi)^n$ in F are identical since we see clearly that up to a factor λ^n these two operators can be written

$$1 + \phi u_n \quad \text{and} \quad 1 + u_n \phi$$

for some $u_n \in L(E, F)$. The identity of the spectral subspaces of ϕu and $u\phi$ follows.

COROLLARY *Let u be a bounded operator in a quasi-complete LCTVS E or more generally suppose there exists a neighborhood V of 0 and of a bounded disk B in E such that $u(V) \subset B$ and that the space E_B generated by B with the norm gauge of B is complete. Then the spectrum of u is a compact subset of the field of scalars.*

Proof: u can be considered as a continuous operator from E into E_B, therefore, by Proposition 4, except for zero, its spectrum is $L(E)$ is identical to the spectrum of a continuous operator in the Banach space E_B which is compact from Proposition 3, Corollary. This ends the proof (if 0 does not belong to the spectrum of u in $L(E)$, u is an isomorphism from E onto E, B is a bounded neighborhood of zero so that E is isomorphic to E_B and we can apply Proposition 3, Corollary directly).

EXERCISE 1 Let M be a locally compact space, let $E = C(M)$. If $f \in E$, call u_f the operator

$$g \mapsto fg$$

in E. Show that the spectrum of u_f is identical to $f(M)$. Conclude that every non-empty subset of the field of scalars can be considered as the spectrum of a continuous operator in some LCTVS; in the case where the given subset is the union of a sequence of compact sets it is even the spectrum of a continuous operator in some Frechet space; in the case where the given subset is a compact subset, it is the spectrum of a continuous operator in a Banach space E (choose

$$E = C(M),$$

M is the compact subset under consideration).

EXERCISE 2   Let E be a complex Banach space, let $u \in L(E)$; consider the function

$$\phi(\lambda) = (\lambda 1 - u)^{-1}$$

defined in the complement of the spectrum of u with values in $L(E)$. Show that $\phi(\lambda)$ tends to 0 when $|\lambda|$ tends to infinity and that for every continuous linear form w on $L(E)$, $\langle \phi(\lambda), w \rangle$ is a holomorphic function of λ. Conclude that the spectrum of u is not empty (if not, apply Liouville's theorem to the entire function zero at infinity $\langle \phi(\lambda), w \rangle$).

EXERCISE 3   Let E be the space of continuous functions f on $\mathbf{R}$ such that

$$\lim_{t \to -\infty} f(t) e^t = 0.$$

Equip E with a natural topology so as to make it a Frechet space. Let u be the translation operator $uf(t) = f(t - h)$ where $h > 0$. Show that u is a continuous operator whose spectrum is empty (show that the series $\sum_{0}^{\infty} \lambda^n u^n$ converges in $L_b(E)$ for any λ). Let u be a *bounded* operator in a LCTVS E, show that the spectrum of u is not empty (use Exercise 2).

EXERCISE 4   Let u be a linear operator in a vector space E. With the notations of Proposition 3, show that the spectral subspaces E_λ are linearly independent.

4   The Riesz theory of compact operators

LEMMA   *Let u be a compact operator in an LCTVS E, let $v = 1 + u$. Then the sequence of kernels of v^n ends up being stationary.*

Proof   Let V be a neighborhood of 0 in E and A a compact disk such that $u(V) \subset A$. Then the space E generated by A with the norm gauge of A is a Banach space, the identity mapping ϕ from E_A into E is compact and u defines a continuous linear mapping u_0 from E into E_A; on the other hand the operator $u_0\,\phi$ induced by u in E_A is compact since u_0 is compact. By Proposition 5, in order to prove the lemma it suffices to prove it for the compact operator $u_0\phi$ in the Banach space E_A. We can therefore suppose that E is a Banach space. We proceed by contradiction; if the sequence of kernels E_n of v^n were strictly increasing we could find an infinite sequence (y_n) such that $u_n \in E_{n+1}$, $\| y_n \| \leqslant 1$, the distance from y_n to E_n being at least equal to $\frac{1}{2}$ (see Chapter 1. Section 13, Lemma). We would have then for $m > n$:

$$uy_n - un_m = (1 - v)y_m - (1 - v)y_n = y_m - x$$

where
$$x = vy_m + (1 - v)y_n \in E_m,$$

so that
$$\| uy_n - uy_m \| \geqslant \tfrac{1}{2}.$$

Now the sequence (uy_n) is relatively compact as it is the image of a sequence extracted from the unit ball and should have a (cluster point), which is absurd. This ends the proof.

Applying the lemma to $v' = 1 + u'$ in E' with the topology of uniform convergence on the compact disks (for which u' is a compact operator by Proposition 1), we see that the sequence of kernels of v'^n ends up being stationary. Now $v'^n = (v^n)'$ and we can write

$$v^n = (1 + u)^n = 1 + u_n,$$

where
$$u_n = nu + \frac{n(n - 1)}{2}\, u^2 + \ldots + u^n$$

is a compact operator in E. Thus (Theorem 3, Corollary 3) v^n is a homomorphism from E onto a closed subspace of finite codimension of E which is therefore the orthogonal of the kernel of $(v^n)'$. Since this last one remains constant for n sufficiently large it follows that the sequence of $v^n(E)$ ends up being stationary and its intersection is a closed subspace F_∞ of finite codimension. Applying Proposition 2, 3) we see that the sequences $v^{-n}(0)$ and $v^n(E)$ are stationary starting from the same rank v and that E is a direct (algebraic) sum of

$$E_\infty = v^{-v}(0) \quad \text{and} \quad F_\infty = v^v(E).$$

It is even their topological direct sum, F_∞ being closed and of finite codimension (Chapter 1, Section 12, Theorem 7, Corollary 3). Furthermore v is nilpotent in E_∞ and is in F_∞ an automorphism in the algebraic sense (Proposition 2, 3), i.e. a bijective mapping from F_∞ onto F_∞. Since it is furthermore a homomorphism in F_∞ by Theorem 2, Corollary 2 applied to the operator $v = 1 + u$ induced by v on F_∞, we see that it is even an isomorphism from F_∞ onto itself. Summing up, we can now state the fundamental theorem of the Riesz theory:

THEOREM 5   *Let u be a compact operator in an LCTVS E, let $v = 1 + u$. Then v is a homomorphism from E onto a closed subspace of E of finite codimension equal to the dimension of the kernel. Let*

$$E_\infty = \bigcup_n v^{-n}(0), \quad F_\infty = \bigcap_n v^n(E),$$

then E_∞ and F_∞ are topological supplements stable under u and v, E_∞ finite dimensional. Then v induces a nilpotent operator in E_∞ and a TVS automorphism in F_∞.

(The first assertion of the theorem is contained in Theorem 2, Corollary 3, and in Proposition 2, Corollary.)

COROLLARY 1   *The image $v(E)$ is the orthogonal of the kernel of v' and the image $v'(E')$ is the orthogonal of the kernel of v.*

The first assertion follows from the fact that $v(E)$ is closed, the second one from the fact that $v'(E')$ is closed in E' equipped with the topology of uniform convergence on compact disks (by Proposition 1, which allows Theorem 3 to be applied to $v' = 1 + u'$), hence also in E' weak.

COROLLARY 2   *The following conditions are equivalent:*

a) *v is bijective*

b) *v is onto*

c) *v is an automorphism*

a') *v' is bijective*

b') *v' is onto*

c') *v' is an automorphism*

(v' *considered as an operator on E' with the topology of uniform convergence on compact disks*).

The equivalence of a) and b) follows from the identity between the dimension of the kernel and the codimension of the image of v; furthermore, v being a homomorphism, a) and b) imply c) and are thus equivalent to it. This also proves the equivalence of a'), b'), c'); finally, the equivalence of b) and a') follows from Corollary 1 which implies, more generally,

COROLLARY 3   *The dimension of the kernel of v, the dimension of the* *kernel of v', the codimension of the image of v and finally the codimension of the image of v' are equal.*

COROLLARY 4   *Let*

$$E'_\infty = \bigcup_n v'^{-n}(0), F'_\infty = \bigcap_n v'^n(E').$$

Then E'_∞ is the orthogonal of F_∞ and F'_∞ is the orthogonal of E_∞. In particular, the dimensions of E_∞ and E'_∞ are the same.

Introducing the integer v at which the four sequences $(v^{-n}(0))$, etc. become constant, it suffices to apply Corollary 1 to $v^n = 1 + u_n$ (where u_n is a compact operator in E).

PROPOSITION 6   *Let u be a compact operator in an LCTVS E. For every $\lambda \neq 0$ of the spectrum of u there exists a non-zero eigenvector, i.e. λ is eigenvalue of u. The corresponding spectral subspace E_λ of E is of finite dimension, equal to the dimension of the spectral subspace E'_λ corresponding to the transpose u' of u; u and u' have the same eigenvalues with the same multiplicities.*

It suffices to apply the preceding corollaries to the compact operator $-(1/\lambda)u$ and to the corresponding operator

$$1 - \frac{1}{\lambda} u = \frac{1}{\lambda} (\lambda 1 - u).$$

COROLLARY   *The subspaces $E_\lambda (\lambda \neq 0)$ are topologically free.*

By this we mean that for every $\lambda \neq 0$ there exists a closed vector subspace of E containing the $E_{\lambda'}$, (λ' different from λ and from 0) and whose intersection with E_λ is zero. Now it suffices to choose F_λ with the notations of Proposition 3.

THEOREM 4   *Let u be a compact operator in an LCTVS E. Then the spectrum of u is a compact set, and every non-zero point of the spectrum is isolated.*

Equivalent statement   The spectrum of u is either finite or is formed of 0 and of points of a sequence that converges to 0.

Proof   We know that the spectrum of u is compact (Proposition 5, Corollary); now let λ be a non-zero element of the spectrum, then we shall show that it is an isolated point. With the notations of Proposition 3, E is a topological direct sum of the closed subspaces E_λ and F_λ stable under u. If u_1 and u_2 are the operators induced by u we clearly have $\sigma(u) = \sigma(u_1) \cup \sigma(u_2)$. Now $\sigma(u_1)$ is reduced to λ and $\sigma(u_2)$ is compact (see above) and does not contain λ (otherwise F would contain a non-zero eigenvector corresponding to λ) whence the conclusion follows.

EXERCISE 1   Let E be an LCTVS.

a) If u is a compact (resp. bounded) linear operator in E and if $v = 1 + u$ is invertible in $L(E)$, show that the inverse of v has the same form of v (use the fact that compact resp. bounded operators form a bilateral ideal in $L(E)$).

b) Conclude that this is equivalent to say that v is invertible in $L(E)$ or in the algebra $L(E_0)$ of continuous linear operators for a locally convex topology on E which is finer than the given topology, having the same bounded subsets, and such that $L(E_0) \subset L(E)$.

c) Let u be a compact operator in E; let E_0 be the space E with a locally convex topology such that every compact operator in E is continuous in E_0. Show that it is equivalent to say that $1 + u$ is invertible in $L(E)$ or in $L(E_0)$.

EXERCISE 2   Let (λ_i) be a sequence tending to zero, let E be the space l^p or the space $c_0(1 \leqslant p \leqslant +\infty)$, u the operator of multiplication by (λ_i) in E. Show that the spectrum of u is composed of 0 and the set (λ_i). Conclude that every non-empty compact subset of the field of scalars where every non-zero point is isolated, is the spectrum of some compact operator in E. Show that Proposition 6 and its Corollaries are false if we do not exclude the spectral value 0.

EXERCISE 3
a) Develop the notion of holomorphic, or meromorphic function, on an open set of the complex plane with values in an LCTVS E.

b) Show that if u is a continuous linear operator in a complex Banach space E, the function $\lambda \mapsto (\lambda 1 - u)^{-1}$ defined in the complement of the spectrum of u with values in the Banach space $L(E)$, is holomorphic. Generalize to the case where u is a bounded operator in an LCTVS E, $L(E)$ with the topology of bounded convergence (see Section 3, Exercise 2).

c) Let u be a compact operator in the complex LCTVS E. Show that the function $(1 - zu)^{-1}$ with values in $L_b(E)$ is meromorphic in all of the complex plane. (It suffices by b) to prove the fact of being meromorphic in a point z such that $1/z$ is an eigen value; for this use the spectral decomposition $E = E_{1/z} + F_{1/z}$.)

PART 3 GENERAL CRITERIA OF COMPACTNESS

1 Šmulian's theorem

PROPOSITION 1   *Let E be a compact space, F a metric space, $\mathfrak{S}$ a set of subsets of E covering E, (f_i) a relatively compact sequence in $C_{\mathfrak{S}}(E, F)$. Then there exists a subsequence that converges in $C_{\mathfrak{S}}(E, F)$.*

Let A be the compact closure of the set of f_i. On A, $\mathfrak{S}$-convergence is equal to pointwise convergence and even equal to pointwise convergence in a dense subset of E, therefore it suffices to find a subsequence of (f_i) which converges on each point of E or even of a dense subset. This leads us to the case where $\mathfrak{S}$ is the set of one-point subsets of E. If E is metrisable and therefore admits a dense subsequence (x_i) since it is compact, A is metrisable for pointwise convergence in the set of (x_i), whence the conclusion in this case. If we do not suppose E metrisable, we consider the compact space $\tilde{E}$ the quotient of E by the equivalence relation "$f_i(x) = f_i(y)$ for every i", then $C_s(\tilde{E}, F)$ can be identified with a closed subspace of $C_s(E, F)$. On the other hand, $\tilde{E}$ is metrisable since its topology is by reasons of compactness the coarsest for which the maps f_i into the metric space F are continuous. We are thus reduced to the former case.

COROLLARY 1   *In Proposition 1, instead of supposing E compact, it suffices to suppose that there exists a sequence of compact subsets of E whose union is dense.*

In fact, it follows from Proposition 1 that we can extract from (f_i) a subsequence which converges on every point of E_1, and from this one a subsequence which converges on every point of E_2, etc., finally, the diagonal procedure gives us a subsequence extracted from (f_i) which converges in every point of $\cup E_i$, which ends the proof since this last set is dense.

COROLLARY 2   *Let E be a Hausdorff LCTVS such that there exists in E' a sequence of weakly compact subsets whose union is total. Then from every relatively compact subset of E we can extract a convergent subsequence.*

We can clearly suppose that the union of weakly compact subsets of E' is even weakly dense. Interpreting as usual E as a space of continuous scalar functions on E' weak with a $\mathfrak{S}$-topology it suffices to apply Corollary 1. In particular suppose that there exists a *sequence* of neighborhoods of 0 in E for a topology compatible with the duality (E, E'), i.e. neighborhoods for $\tau(E, E')$, whose intersection is 0; we see immediately by polarity that there exists in E' a sequence of weakly compact disks whose union is dense. In this case we can apply Corollary 2. In particular, if E is the weak space associated with a metrisable LCTVS we find

THEOREM 1 (ŠMULIAN) *Let E be a metrisable LCTVS, then from every weakly relatively compact sequence of E, we can extract a weakly convergent subsequence.*

EXERCISE 1 Show that Proposition 1 is false if we do not suppose F metrisable (choose E reduced to a point).

EXERCISE 2 Let E be a compact space, F a metric space, A a subset of $C_s(E, F)$, f a point in the closure of A. Show that f is in the closure of a countable subset of A. (For every integer $n > 0$, show that there exists a finite subset F_n of A such that for every $x = (x_1, \ldots, x_n) \in E^n$ there exists $f_x \in F_n$ such that

$$d(f_x(x_i), f(x_i)) \leqslant 1/n$$

for $1 \leqslant i \leqslant n$.)

EXERCISE 3 Let E be a separable LCTVS then the weakly compact subsets of the dual are metrisable, therefore from every weakly relatively compact sequence in E we can extract a weakly convergent subsequence. Show that this statement is false if E is not separable (Example: $E = l^\infty$ and the sequence of coordinate forms on E).

2 Eberlein's theorem

PROPOSITION 2 *Let E be a compact space, F a metric space, $\mathfrak{G}$ a set of subsets of E covering E, A a subset of $C_\mathfrak{G}(E, F)$. Then A is relatively compact if and only if every sequence in A admits a limit point in $C_\mathfrak{G}(E, F)$.*

We need only prove the sufficiency; as the condition stated implies that A is precompact (Weil's criterion, Chapter 0) it suffices to show that the closure of A in $C_\mathfrak{G}(E, F)$ is complete. For this we show that its closure for pointwise convergence is complete for pointwise convergence and, a fortiori, for $\mathfrak{G}$-convergence (Chapter 0, Section 1, Proposition 6, Corollary) which leads us back to the case of pointwise convergence. For every $x \in E$, $A(x)$ is a subset of E such that every subsequence admits a cluster point, therefore $A(x)$ is relatively compact in F since F is metric. By Tychonoff, A is relatively compact in $\mathscr{F}_s(E, F) = F^E$ with pointwise convergence. It remains to be shown that the closure of A in $\mathscr{F}_s(E, F)$ is contained in $C_s(E, F)$, therefore that a mapping f from E into F in the closure of A for pointwise convergence is continuous. We proceed by contradiction; if this were not so there would exist an

$a \in E$ where f is not continuous, therefore an $\varepsilon > 0$ such that for every neighborhood V of a an $x \in V$ would exist with

$$d(f(x), f(a)) \geqslant \varepsilon.$$

We could construct by induction a sequence (f_i) extracted from A and a sequence (x_i) extracted from E, with $x_0 = a$ such that

1) $d(f_n(x_i), f(x_i)) \leqslant 1/n$ for $0 \leqslant i \leqslant n - 1$,

2) $d(f_i(x_n), f_i(x_0)) \leqslant 1/n$ for $0 \leqslant i \leqslant n$,

3) $d(f(x_n), f(x_0)) \geqslant \varepsilon.$

Suppose the 2 sequences constructed up to rank $n - 1$, then determine f_n so as to obtain 1) which is possible because f is in the closure of A, then construct x_n so as to satisfy 2) and 3) using the fact that the set of x_n of E satisfying 2) is a neighborhood of $x_0 = a$. Let g be a cluster point of the sequence (f_n) in $C_s(E, F)$ (which exists by hypothesis on A) and x a cluster point of (x_n) in E. We have $f(x_i) = g(x_i)$ for every i by 1), $f_i(x) = f_i(x_0)$ for every i by 2), and, taking 1) into consideration

$$d(f_n(x), f(x_0)) \leqslant 1/n$$

for every n, whence $g(x) = f(x_0)$. Now g being a continuous function, $g(x)$ is a cluster point of $(g(x_i))$, therefore $f(x_0)$ is a point cluster of $(f(x_i))$, which contradicts 3) and ends the proof.

THEOREM 2 (EBERLEIN)   *Let E be an LCTVS, A a subset of E. Suppose that every sequence extracted from A admits a cluster point and that the closed convex hull of A is complete for $\tau(E, E')$. Then A is relatively compact.*

Since A is precompact, it suffices to show that its closure is complete and a fortiori that its weak closure is weakly complete, which reduces the problem to the case where E has the weak topology. Since $\bar{\Gamma}(A)$ is complete for $\tau(E, E')$, it will be closed in the completion $\hat{E}$ of E for $\tau(E, E')$, therefore weakly closed in this space since it is a convex subset; it suffices then to show that the weak closure of A in $\hat{E}$ is weakly compact since this closure is contained in E. Thus we may assume that E is complete for $\tau(E, E')$. Since A is clearly bounded hence relatively compact in the algebraic dual of E' with the weak topology, it suffices to show that every form X on E' in the closure of A for convergence is in E. Now for every weakly compact subset K of E' the restriction X_K of X to K is in the closure (pointwise convergence) of the set A_K of restrictions of $x \in A$ to K. It is clear that every sequence extracted from A_K has a cluster point in $C_s(K)$, then by Proposition 2, A_K is relatively compact in $C_s(K)$, and X_K is continuous. Since this is true for every weakly compact disk K in E' this implies that x is in E since E is complete for $\tau(E, E')$ (Chapter 2, Section 14, Theorem 10, Corollary 3). Applying Theorem 2 for a weak topology we obtain the following result (equivalent to one already known).

COROLLARY   *Let E be a quasi-complete LCTVS. A subset of E is weakly relatively compact if and only if every subsequence admits a weak cluster point.*

THEOREM 3   *Let K be a compact space, A a subset of $C(K)$. Then A is weakly relatively compact if and only if A is bounded and relatively compact in $C_s(K)$.*

The necessity is trivial. For the sufficiency, by virtue of the theorem on weak closure in $C(K)$ or by virtue of Section 1, Proposition 1, we can extract a sequence converging simply to a continuous function. The conclusion follows from

PROPOSITION 3   *A sequence (f_i) in $C(K)$ (K compact), converges weakly to an $f \in C(K)$ if and only if it is bounded and converges pointwise to f.*

The necessity is trivial. Since the continuous linear forms on $C(K)$ are the measures on K, the sufficiency follows from the Lebesgue theorem.

EXERCISE 1   We shall say that a space has the property (E) if every subset with every sequence extracted from it admitting a cluster point is relatively compact. Let E be a locally compact or metrisable space, $\mathfrak{S}$ a set of subsets of E covering E, F a uniform space. Show that $C_\mathfrak{S}(E, F)$ has the property (E) if and only if F has it. (Examine the case where E is compact, then the case of pointwise convergence, then the case where F is the field of reals, viewing the topology of F as the coarsest for which a certain family of real functions on F is continuous.)

EXERCISE 2   Let E be an LCTVS, A a convex subset of E.

a) A is weakly relatively compact if and only if $\bar{A}$ is complete and its image by every continuous linear mapping from E into an arbitrary Banach space F is weakly relatively compact. (Notice that E is isomorphic to a topological vector subspace of a product of Banach spaces.)

b) Show that in this statement we can choose F to be the space l^∞ of bounded sequences. (We can suppose by a) that E is a Banach space and we can furthermore suppose A disked. We must show that the identity mapping from $G = E_A$ into E is weakly compact or equivalently that the transpose mapping maps the unit ball of E' into a subset of G' relatively compact for $\sigma(G', G'')$ (Chapter 2, Section 18, Theorem 13, Corollary 3). Show that it suffices to show that the image of every sequence (x_i') extracted from the unit ball of E' is relatively compact for $\sigma(G', G'')$ and examine the mapping $ux = (\langle x, x_i \rangle)$ from E into l^∞.

c) Suppose E is separable. Then on statement a) we can suppose that F is the space c_0. (With the notations of b), notice that it suffices to prove that the mapping $E' \rightarrow G'$ has a restriction to the unit ball of E' which is continuous for $\sigma(E', E)$ and $\sigma(G', G'')$ and only continuous at the origin, and that the unit ball of E' being weakly metrisable it suffices to show that for every sequence (x_i') in E' tending to zero for $\sigma(E' E)$, its image in G tends to zero for $\sigma(G', G'')$). The hypothesis that E is separable is essential, if not, a subset A of E may be not weakly relatively compact, its image in C_0 by every continuous linear mapping is *compact*.

EXAMPLE $E = l^\infty$, A a subset of l^∞ which is bounded, weakly metrisable and not weakly relatively compact, for example a weak Cauchy sequence not weakly convergent (see Section 3, Supplementary Exercise 4).

EXERCISE 3   Prove the analogue of Exercise 2 replacing weak compactness by compactness (the proof is similar but there is no need to use Eberlein's theorem).

EXERCISE 4   Let E be an LCTVS, A a convex subset of E. Then A is weakly relatively compact if and only if its closure is complete, and for every decreasing sequence of convex subsets A_i of A we have $\bigcap\limits_i \bar{A}_i \neq \phi$ (By Exercise 2 a) reduce to the case where E is a Banach space, then it suffices to show that every sequence (x_i) extracted from A has a weak cluster point. This brings us to the case where E is separable hence, where there exists a weakly dense sequence (x_i') in E'. Extracting if needed from (x_i) a partial subsequence, we can suppose that for every j,

$$\lim \langle x_i, x_j' \rangle$$

exists. Let A_n be the convex hull of the set of x_i with $i \geqslant n$, let

$$x \varepsilon \bigcap_n \bar{A}_n,$$

show that x_i tends weakly to x using Chapter 2, Section 18, Exercise 1.)

EXERCISE 5   Let E be a completely regular space whose topology T

is finer than a certain metrisable topology T_0. Let A be a subset of E. The following conditions are equivalent:

a) A is relatively compact.

b) Every sequence extracted from A admits a cluster point.

c) For every sequence extracted from A, there exists a subsequence extracted from the latter which is convergent. (Show that b) implies c) viewing the topology T as the l.u.b. of a family of metrisable topologies T_i finer than T_0. Show then that a) implies that the closure of A is metrisable).

EXERCISE 6   Let K be a compact space, A and B two weakly relatively compact subsets of $C(K)$. Show that the set AB of $fg(f \in A, g \in B)$ is weakly relatively compact. (Use Theorem 3.) For a deeper and more general result, see Part 4, Section 2, Corollary 1.

EXERCISE 7   Let M be a locally compact space, (f_i) a sequence in $C_0(M)$. It is a weak Cauchy sequence if and only if it is bounded and converges at each point of M. (Use the Lebesgue theorem.)

3 Krein's theorem

THEOREM 4 (KREIN)   *Let E be an LCTVS, A a weakly compact subset of E. Then its closed convex hull $\bar{\Gamma}(A)$ is weakly compact if and only if it is complete for the given topology.*

Applying the same statement to E with $\tau(E, E')$ (which does not change the corresponding weak topology) we see that it suffices that $\bar{\Gamma}(A)$ be complete for $\tau(E, E')$ which is a priori less restrictive.

COROLLARY 1   *Let E be a quasi-complete LCTVS, then in E the closed convex hull of a weakly compact subset is weakly compact.*

Theorem 4 is contained in the following statement which appears more general.

COROLLARY 2   *Let E be an LCTVS, A a compact subset of E. Then its closed convex hull $\bar{\Gamma}(A)$ is compact if and only if it is complete for $\tau(E, E')$.*

(Theorem 4 can be obtained by applying this statement to a space E with the weak topology.)

The condition is necessary since a compact subset of E is complete therefore complete for $\tau(E, E')$. Conversely, suppose $\bar{\Gamma}(A)$ complete for $\tau(E, E')$. Since $\bar{\Gamma}(A)$ is precompact for the given topology it suffices to show that it is complete for this topology and a fortiori to show that it is weakly compact. Thus we are reduced the case where the given topology of E is the weak topology, and we may, as in the proof of Theorem 2, assume E to be complete for $\tau(E, E')$. It will suffice to show that the disked hull B of A is weakly relatively compact, i.e. that the identity mapping u from the normed space $E_B = F$ generated by B (with the norm gauge of B) in E is weakly compact. By Chapter 2, Section 18, Theorem 13, Corollary 1, this means that u' transforms the equicontinuous subsets A' of E' into relatively compact subsets of F for $\sigma(F', F'')$. Now consider the subspace H of elements of F' whose restrictions to A are continuous for $\sigma(E, E')$; it is clearly a closed, therefore complete vector subspace of F' and the norm induced on H by F' is also that induced by the space $C(A)$ of continuous functions on the compact A with $\sigma(E, E')$ (B is the disked hull of A). Therefore, H is a closed subspace of $C(A)$, hence weakly closed, and for a subset H to be relatively compact in F' for $\sigma(F', F'')$ it is (necessary and) sufficient that it be weakly relatively compact in $C(A)$, then by continuity it will be weakly relatively compact in the Banach space F', i.e. for $\sigma(F', F'')$. Now clearly $u'(E') \subset H$ and we must show only that u' transforms an equicontinuous subset of E' into a weakly relatively compact subset of $C(A)$. Then u' transforms $x' \in E'$ into the restriction of x' to A and we must show that if x' runs through an equicontinuous subset of E', its restriction to A runs through a weakly relatively compact set in $C(A)$. Now, by Theorem 3, this is clear (since $u' : E' \to C(A)$ being continuous from E' weak into $C_s(A)$ transforms an equicontinuous subset, therefore weakly relatively compact subset of E', into a relatively compact subset of $C_s(A)$, and we verify besides that it is bounded. This ends the proof (but notice that in the form of Theorem 3 the proof uses essentially Eberlein's theorem and the Lebesgue theorem in integration).

PROPOSITION 4   *Let K be compact with a measure μ, f a weakly continuous mapping from K into a quasi-complete LCTVS E. Then f is weakly integrable in E.*

Recall that this means that on one hand f is scalarly integrable (trivial, since scalarly continuous) and furthermore the linear form x on E' given by

$$\langle x, x' \rangle = \int \langle f(t), x' \rangle \, d\mu$$

belongs to E. Now, supposing $\| \mu \| \leqslant 1$, which is permitted, we see that x belongs to the polar, in the algebraic dual of E', of the polar of

$f(K)$, that is, to the bipolar of $f(K)$, that is to the weakly closed disked hull of $f(K)$ in the weak completion of E. It suffices then to show that this hull is contained in E, therefore that the weakly closed disked hull of $f(K)$ in E is already weakly compact. Now $f(K)$ being weakly compact by continuity, it suffices to apply Theorem 4, Corollary 1, which is also valid for the closed disked hull, either as a result of the given proof or as an immediate corollary of the given statement.

EXERCISE 1   Show that the conclusion of Proposition 4 remains valid if we suppose K locally compact only, the measure μ bounded, f weakly continuous and *bounded*. (Examine first the case where μ has a compact support, by Proposition 4, then pass to the limit for any given μ).

EXERCISE 2   Let K be a compact subset of a complete LCTVS. Show that the closed convex hull of K is the set of weak integrals

$$\int_K x \, d\mu(x)$$

where μ runs through the set of positive measures of norm 1 on K. Show that Theorem 4 would be an easy consequence of Proposition 4.

Supplementary exercises

Here are some exercises on compactness which could have been given starting from Chapter 2, with the exception of Exercise 3, 3) and Exercises 4 and 5 which use Šmulian's theorem.

EXERCISE 1   Let (x_i) be a Cauchy sequence in an LCTVS E, show that its closed disked hull A is metrisable. (Work out the case where E is complete and Hausdorff, then (x_i) convergent, then the case where the limit is zero noticing that the sum of two metrisable compact sets A and B is a homeomorphic compact set since it is metrisable to a separable quotient of $A \times B$. Then using Chapter 2, Section 13. Exercise 2, notice that A is isomorphic to a quotient space of the unit ball of l^1 with $\sigma(l^1, c_0)$ and that the latter is a metrisable compact set.)

EXERCISE 2.   Let E be an LCTVS, let $\mathfrak{G}$ be the set of subsets A of E such that from every sequence extracted from A, we can extract a Cauchy subsequence ($\mathfrak{G}$ then contains the metrisable precompact spaces).

1) Let u be a linear mapping from E into an LCTVS F. Show that if u transforms convergent sequences into convergent sequences, then u transforms Cauchy sequences into Cauchy sequences and the $A \in \mathfrak{G}$ into precompact subsets. (For the first part, let A be the closed disked

hull of the Cauchy sequence (x_i), show that the restriction of u to A is uniformly continuous, noticing that by Chapter 2, Section 14, Lemma, it suffices to show the continuity of this restriction at the origin and that it then suffices, by Exercise 1, to show this continuity for the sequences in A tending to 0. The second point follows from the first in a purely topological way, using Weil's criterion.)

2) Let A' be a set of linear forms on E which are continuous on the sequences, then the following conditions on A' are equivalent:

    a) A is precompact for uniform convergence on the convergent sequences of E;

    b) A' is precompact for $\mathfrak{S}$-convergence;

    c) every convergent sequence (or also: every Cauchy sequence) in E, converges uniformly on A'. (Show that this statement is equivalent to the first one, using Chapter 2, Section 13, Theorem 12.)

Exercise 3   Let E be LCTVS, $\mathfrak{S}$ the set of subsets A of E such that from every sequence extracted from A we can extract a weak Cauchy subsequence ($\mathfrak{S}$ contains then the bounded sets which are metrisable for the weak uniform structure since "bounded" is equivalent to "weakly precompact").

1) Let u be a linear mapping from E into an LCTVS F. If u transforms the weakly convergent sequences into convergent sequences, it transforms weak Cauchy sequences into Cauchy sequences, and the $A \in \mathfrak{S}$ into precompact subsets. (Particular cases of the preceding exercise applied to E weak; the same proof for 2).)

2) Let A' be a subset of E'. The following conditions are equivalent:

    a) A' is precompact for uniform convergence on the weakly convergent sequences of E;

    b) A' is precompact for $\mathfrak{S}$-convergence;

    c) every weakly convergent sequence (or also: every weak Cauchy sequence) in E, converges uniformly on A'.

Following G. Köthe we call such a subset of E' *limited* ("begrenzt"); this notion depends only on the dual system (E, E') (so that the limited subsets of E are defined at the same time).

3) $A' \subset E'$ is limited if it is precompact for $\tau(E', E)$. This condition is also necessary if E is separable or metrisable (more generally, each time that Šmulian's theorem is valid in E, as it means that the weakly compact subsets of E are in $\mathfrak{S}$!).

4) A weakly continuous linear mapping (only the dual systems are

pertinent here) from one LCTVS into another transforms limited subsets into limited subsets. Show that for a subset A of E, the following conditions are equivalent:

a) A is limited;

b) every weakly continuous linear mapping from E into a *separable* LCTVS F transform A into a precompact subset;

c) every weakly continuous linear mapping from E into c_0 transforms A into a precompact subset of c_0.

(a) $\Rightarrow$ b) follows from 3), b) $\Rightarrow$ c) is trivial and c) $\Rightarrow$ a) can be seen using a canonical bijective correspondence between weakly continuous linear mappings from E into c_0, and sequences weakly convergent to 0 in E'). Compare with Section 2, Exercise 2, c).

EXERCISE 4   Let (E, E') be a dual system, $\mathfrak{G}$ the set of subsets A of E such that from every sequence extracted from A we can extract a weak Cauchy subsequence, A' the set of analogous subsets of E', $\mathscr{L}$ (resp. $\mathscr{L}'$) the set of limited subsets of E (resp. E') (see Exercise 3).

1) Show that the following conditions are equivalent:

a) $\mathfrak{G} \subset \mathscr{L}$; a') $\mathfrak{G}' \subset \mathscr{L}'$;

b) (resp. c) the weakly convergent sequences (resp. the weak Cauchy sequences) in E are limited;

b') (resp. c') the same statement relative to E';

d) For every sequence (x_i) (resp. (x_i')) in E (resp. E') which tends weakly to 0 we have $\lim \langle x_i, x_i' \rangle = 0$.

2) If E is separable, it cannot have the preceding property unless its bounded subsets are precompact for $\tau(E, E')$; in particular if E is a normable space this E must be finite dimensional. The same conclusion follows if E is a reflexive not necessarily separable Banach space (use Exercise 3, 3)).

3) Let K be a compact Stonian space, set $E = C(K)$, show that (E, E') satisfies the conditions of 1). (Use Part 4, Section 1, Exercise 12, 2) showing that a weakly convergent sequence in E' converges for $\sigma(E, E'')$ and also the same Section 1, Exercise 11, showing that in a dual of a space $C(K)$ a sequence convergent for $\sigma(E', E'')$ converges for $\tau(E', E)$.)

4) Noticing that if in 3) K is infinite (for example $C(K) = l^\infty$ when K is the Stone compactification of the integers), there exist weak

P

Cauchy sequences not weakly convergent in $E = C(K)$ (see for example Part 4, Section 2, Exercise 3, 2º), conclude from this that there exist in E limited subsets that are not even weakly relatively compact. (Compare with Section 2, Exercises 2 and 3.)

EXERCISE 5

1) Let μ be a positive measure countable at infinity on a locally compact space M, let $E = L^1(\mu)$. Show that from every bounded sequence of the dual E' we can extract a weakly convergent subsequence. (Reduce to the case where μ is bounded replacing μ by an equivalent measure which replaces $L^1(\mu)$ by an isomorphic space; in this case we have $L^\infty \subset L^1$, and the inclusion mapping is continuous for $\sigma(L^\infty, L^1)$ and the weak topology $\sigma(L^1, L^\infty)$ of L^1, therefore the unit ball of L^∞ is a weakly compact subset of L^1. Then apply Šmulian's theorem in L^1.)

2) Let μ be a measure on a locally compact space. Show that the limited subsets of $E = L^1(\mu)$ (see Exercise 3) are identical to the precompact subsets of E. (In order to show that a limited subset is precompact we can suppose that it is a sequence in $E = L^1(\mu)$; reduce to the case where μ is countable at infinity; in this case, the assertion follows trivially from 1).)

EXERCISE 6   Let E be an LCTVS whose strong dual is separable, A linear mapping u from E into an LCTVS F, transforming weakly convergent sequences into convergent sequences, transforms bounded subsets into precompact subsets—and conversely. (Use Exercise 3, 1), noticing that the bounded subsets of E are weakly metrisable in this case.)

PART 4   WEAK COMPACTNESS IN L¹

1 The Dunford–Pettis criterion and its first consequences

In what follows, M stands for a locally compact space, μ a positive measure on M. We suppose the general theory of integration (Bourbaki, *Integration*, Chapters 1 to 5) is known and we follow in general Bourbaki's terminology. Recall that $L^1(\mu) = L^1$ the Banach space of equivalent classes of summable functions, with the norm

$$\| f \| = \int | f(s) | \, d\mu \, (s),$$

and that its dual can be identified with $L^\infty(\mu) = L^\infty$, the space of classes of measurable and bounded functions with the norm

$$||f||_\infty = \text{l.u.b. in measure } |f(s)|.$$

The coupling is given by

$$\langle f, g \rangle = \int fg \, d\mu \; (f \in L^1, g \in L^\infty).$$

It is sometimes convenient to introduce the space $\mathscr{M}(\mu) = \mathscr{M}$ of all classes of measurable scalar functions (modulo the equality locally almost everywhere), with the topology *of convergence in measure on every compact set*, where a fundamental system of neighborhoods of 0 is formed by sets $V(K, \varepsilon)$, K compact $\subset M$ and $\varepsilon > 0$ formed by the f such that the set of points of K, where $|f(s)| \geqslant \varepsilon$, is of measure $> \varepsilon$. (We thus obtain a TVS topology but in general not a LCTVS). The convergent sequences for this topology are said to be *sequences convergent in measure on every compact set*. The L^p spaces can be imbedded canonically and continuously in $\mathscr{M}$ (which explains the use of $\mathscr{M}$). Recall *Egoroff's theorem*, a sequence (f_i) which converges almost everywhere, converges in measure on every compact set.

If A is a subset of M, we call ϕ_A its characteristic function or also its class in L^∞ if A is measurable. With this notation we have for $f \in L^1$:

$$\langle f, \phi_A \rangle = \int_A f \, d\mu.$$

LEMMA 1 *Let (f_n) be a sequence in L^1 such that $(\langle f_n, \phi_A \rangle)$ is convergent for any open set A. Then we have*

a) *For every $\varepsilon > 0$ there exists $\eta > 0$ such that A measurable, $\mu(A) \leqslant \eta$ implies $\langle |f_n|, \phi_A \rangle \leqslant \varepsilon$ for every n.*

b) *For every $\varepsilon > 0$ there exists a compact set $K \subset M$ such that*

$$\langle |f_n|, \phi_{\mathbf{C}K} \rangle \leqslant \varepsilon$$

for every n.

Proof

a) It is well known that for a given $f \in L^1$ we can find for every $\varepsilon > 0$ an $\eta > 0$ such that A measurable, $\mu(A) \leqslant \eta$ implies

$$\langle |f|, \phi_A \rangle \leqslant \varepsilon.$$

(Since, letting M_n be the set of $s \in M$ such that $|f(s)| \geqslant n$, then

$$\cap M_n = \phi,$$

therefore, by Lebesgue's theorem

$$\langle |f|, \phi_{M_n} \rangle \to 0,$$

so that there exists n such that

$$\langle |f|, \phi_{M_n} \rangle \leqslant \varepsilon/2;$$

choose $\eta = (\varepsilon/2n)$, then A is measurable and $\mu(A) \leqslant \eta$ implies

$$\langle |f|, \phi_A \rangle \leqslant \langle |f|, \phi_{A \cap \complement M_n} \rangle + \langle |f|, \phi_{M_n} \rangle \leqslant \eta \mu(A) + \varepsilon/2 \leqslant \varepsilon.$$

Suppose first that for every open set A, $\langle f_n, \phi_A \rangle$ tends to 0. In order to prove a) it suffices to show that we can find $\eta > 0$ such that A measurable, $\mu(A) \leqslant \eta$ implies

$$\langle f_n, \phi_A \rangle \leqslant \varepsilon$$

for every n, since we shall then have (supposing the f_n real)

$$\langle |f|, \phi_A \rangle = \langle f, \phi_{A_1} \rangle - \langle f, \phi_{A_2} \rangle$$

(where A_1, or A_2, is the set of points of A where f_n is positive, resp. negative) whence,

$$\langle |f|, \phi_A \rangle \leqslant \varepsilon + \varepsilon = 2\varepsilon.$$

We proceed now by contradiction supposing that for any $\eta > 0$, there exists a measurable set A and an index n such that

$$\mu(A) < \eta, \ |\langle f_n, \phi_A \rangle| > \varepsilon.$$

We can clearly suppose A open (replacing it by a slightly larger open set and n as big as desired. We shall construct by simultaneous induction a strictly increasing sequence of indices (n_i) and a sequence of open sets A_i, such that we have, where A_{ij} are the set of elements of A_i which do not belong to either A_j or A_k for $k < i$,

(1) $$|\langle f_{n_j}, \phi_{A_{ij}} \rangle| \leqslant 2^{-i-j} \text{ for every } i, j$$

(2) $$|\langle f_{n_j}, \phi_{A_j} \rangle| \geqslant \varepsilon \text{ for every } j.$$

Suppose the construction is done up to rank k; by the remark at the beginning we can find an $\eta > 0$ such that

$$\langle |f_j|, \phi_A \rangle \leqslant 2^{-2k-1} \leqslant 2^{-j-(k+1)}$$

for $$\mu(A) \leqslant \eta, \ j \leqslant k$$

and since (f_n) tends to 0 on the characteristic functions of open sets and therefore on their linear combinations $\phi_{A_{ij}}(i, j \leqslant k)$ we can find an index n such that

$$|\langle f_m, \phi_{A_{ij}} \rangle| \leqslant 2^{-2k-1} \leqslant 2^{-i-(k+1)}$$

for $$m \geqslant n, \ i, j \leqslant k.$$

Finally, by hypothesis we can find an open set $A = A_{k+1}$ and $m = n_{k+1}$ such that

$$\mu(A) \leqslant \eta, \ m \geqslant n, \ |\langle f_m, \phi_A \rangle| \geqslant \varepsilon,$$

and we verify immediately that the hypothesis of induction is still satisfied. Let $A = \cup A_k$, which is open, and since

$$A = A_j \cup \bigcup_i A_{ij} \text{ (disjoint union)}$$

$$\langle f_{n_j}, \phi_A \rangle = \langle f_{n_j}, \phi_{A_j} \rangle + \sum_i \langle f_{n_j}, \phi_{A_{ij}} \rangle$$

then by (1) and (2)

$$|\langle f_{n_j}, \phi_A \rangle| \geqslant \varepsilon - \sum_i 2^{-i-j} = \varepsilon - 2^j.$$

But on the other hand we should have

$$\langle f_{nj}, \phi_A \rangle \to 0$$

which is a contradiction.

We now consider the general case. We claim that there exists an $n > 0$ and an $\eta > 0$ such that $\mu(A) \leqslant \eta$, $p, q \geqslant n$ imply

$$\langle |f_p - f_q|, \phi_A \rangle \leqslant \varepsilon/2.$$

If not, we could find two sequences of indices p_k and q_k and a sequence from A_k such that

$$u(A_k) \leqslant \frac{1}{k}, \quad p_k, q_k \geqslant k,$$

$$\langle |f_{p_k} - f_{q_k}|, \phi_{A_k} \rangle \geqslant \varepsilon/2,$$

which contradicts the preceding result as the sequence

$$(f_{p_k} - f_{q_k})$$

is a sequence which tends to 0 on the ϕ_A, A open. Then let n and η be as above; choosing η sufficiently small we can furthermore suppose that $\mu(A) \leqslant \eta$ implies also

$$\langle |f_m|, \phi_A \rangle \leqslant \varepsilon/2$$

for $m \leqslant n$. We then have for $\mu(A) \leqslant \eta$ and $m \geqslant n$

$$\langle |f_m|, \phi_A \rangle \leqslant \langle |f_n|, \phi_A \rangle + \langle |f_m - f_n|, \phi_A \rangle \leqslant \varepsilon/2 + \varepsilon/2 = \varepsilon$$

whence $\langle |f_m|, \phi_A \rangle \leqslant \varepsilon$

for *every* m which ends the proof of the first part of the lemma.

b) Let μ_n be the measure of density f_n with respect to μ. From the hypothesis that $\mu_n(A)$ is a converging sequence for every open set A, we wish to conclude that there exists a compact set K such that

$$|\mu_n|(\complement K) \leqslant \varepsilon$$

for every n. Interpreting the μ_n as bounded measures on the compact set $\hat{M}$ obtained by the adjunction of "a point at infinity" ω, the hypothesis remains valid in this interpretation of the μ_n, and the conclusion means that we can find a neighborhood V of ω in $\hat{M}$ such that $|\mu_n|(V) \leqslant \varepsilon$ for every n. Now we can find a measure ν on $\hat{M}$ such that the μ_n belong to $L^1(\nu)$ (identified in the usual fashion with a space of measures) for example

$$\nu = \sum_n 2^{-n} \frac{\mu_n}{\|\mu_n\|}.$$

By the first part we can find $\eta > 0$ such that for every measurable subset A of M, $\nu(A) \leqslant \eta$ implies $|\mu_n|(A) \leqslant \varepsilon$. In particular, we can choose a neighborhood V of ω such that

$$\nu(V \cap \complement \omega) \leqslant \eta,$$

which is the desired V.

THEOREM 1 (DUNFORD-PETTIS)   *Let H be a subset of L^1. Then H is weakly relatively compact if and only if it satisfies the two following conditions:*

   a) *For every $\eta > 0$, there exists $\varepsilon > 0$ such that A measurable and*

$$\mu(A) \leqslant \eta$$

implies

$$\langle |f|, \phi_A \rangle \leqslant \varepsilon$$

for every $f \in H$.

   b) *For every $\varepsilon > 0$ there exists a compact set $K \subset M$ such that*

$$\langle |f|, \phi_{\complement K} \rangle \leqslant \varepsilon$$

for every $f \in H$.

   Suppose H weakly relatively compact. If a) were not true there would exist a sequence (A_n) of measurable subsets of M and a sequence (f_n) extracted from H such that

$$\langle |f_n|, \phi_{A_n} \rangle > \varepsilon, \quad \mu(A_n) \leqslant 1/n.$$

But from Šmulian's theorem we can extract from (f_n) a weakly convergent subsequence and this contradicts the first part of the preceding lemma. Likewise, if b) were not true we could find a sequence of compact sets $K_n \subset M$ pairwise disjoint and a sequence (f_n) extracted from H such that $\langle |f_n|, \phi_{K_n} \rangle > \varepsilon$. Let (g_n) be a weakly convergent subsequence of (f_n) (Šmulian's theorem), let K be a compact set such that

$$\langle |g_n|, \phi_{\complement K} \rangle \leqslant \varepsilon/2$$

for every n we would have

$$\langle |g_n|, \phi_{K_m} \rangle \leqslant \frac{\varepsilon}{2} + \langle |g_n|, \phi_{K \cap K_m} \rangle,$$

and the second term at the right tends to 0 for $m \to \infty$ uniformly with respect to n by Lemma 1, a) whence

$$\langle | g_n |, \phi_{K_m} \rangle < \varepsilon$$

for every n for $m \geqslant m_0$, which is absurd.

Suppose a) and b) verified, we shall show that H is weakly compact. We use the

LEMMA 2 *Let H be a subset of a Banach space E such that for every $\varepsilon > 0$ there exists a weakly compact subset H' of E such that every $x \in H$ be at a distance $< \varepsilon$ from H'. Then H is weakly relatively compact.*

Let B be the unit ball of E, E'' the bidual of E, $\bar{B}$ the closure of B in E'' with the weak topology (i.e. $\sigma(E'', E')$). H being clearly bounded, it suffices to show that its weak closure $\bar{H}$ in E'' is contained in E. Now from

$$H \subset H' + \varepsilon B$$

we conclude

$$\bar{H} \subset H' + \varepsilon \bar{B}$$

since H' being weakly compact and $\varepsilon \bar{B}$ weakly closed the second term is weakly closed in E''. A fortiori we shall have

$$H \subset E + \varepsilon \bar{B}$$

for every $\varepsilon > 0$, whence $H \subset E$ since E is strongly closed in E''. The conclusion follows.

We now return to the conditions of Theorem 1. For every compact $K \subset M$, consider the set $H' = \phi_K H$ of products $\phi_K f, f \in H$. By Lemma 2 and condition b) of the theorem it suffices to show that the sets H' are weakly relatively compact, which leads us to the case where all the $f \varepsilon H$ vanish outside a fixed compact set K, therefore to the case where M is itself a compact set K. In this case we have $L^\infty \subset L^1$. We see first of all that a) implies that H is bounded. Let A be the unit ball of L^∞; on A, convergence to 0 in the sense of the topology induced by L^1 implies uniform convergence on H. In fact, by condition a) on H we see easily that even the convergence in measure on A implies uniform convergence on H. Now on L^∞ the topology induced by L^1 is the topology of uniform convergence on the subset A of L^1, and the preceding argument means that for every $\varepsilon > 0$ there exists $\eta > 0$ such that

$$A \cap \eta A^0 \subset \varepsilon H^0.$$

By polarity this can be written

$$\frac{1}{\varepsilon} H \subset \bar{\Gamma}\left(A^0, \frac{1}{\eta} A\right)$$

where A^0 is simply the unit ball B of L^1. Then a fortiori

$$H \subset \frac{\varepsilon}{\eta} A + \varepsilon B,$$

i.e. we have the conditions of Lemma 2, with $H' = \frac{\varepsilon}{\eta} A$. It suffices to notice that A is a weakly compact subset of L^1 which is immediate, the identity mapping from L^∞ into L^1 being continuous for the weak topology $\sigma(L^\infty, L^1)$ of L^∞ and the weak topology of L^1 and transforming then the weakly compact subset A of L^∞ into a weakly compact subset of L^1.

COROLLARY 1   *Let $H \subset L^1$ and let H' be the set of $f \in L^1$ such that there exists $g \in H$ with $|f| \leqslant |g|$. Then H is weakly relatively compact if H' is.*

In particular

COROLLARY 2   *A subset of L^1 which is order bounded is weakly relatively compact.*

(Choose in Corollary 1 H reduced to an element.)

COROLLARY 3   *In a space L^1, a weak Cauchy sequence is weakly convergent.*

By Lemma 1, the two conditions of Theorem 1 are satisfied.

PROPOSITION 1   *Every bounded sequence in L^∞ converging to an f in measure on every compact set, converges to f uniformly on every weakly compact subset of L^1 (i.e. for the Mackey topology $\tau(L^\infty, L^1)$).*

The proof is immediate by conditions a) and b) of Theorem 1. The converse is true, see Exercise 1.

COROLLARY   *Let u be a weakly continuous linear mapping from L^∞ into an LCTVS E. Then u transforms the bounded sequences convergent in measure into convergent sequences of E.*

In fact, the transpose u' of u is a linear mapping from E' into L' continuous for $\sigma(E', E)$ and $\sigma(L^1, L^\infty)$ and thus transforms the weakly compact subsets and a fortiori the weakly closed equicontinuous subsets of E' into weakly compact subsets of L^1. The conclusion of the corollary means precisely that the sequences (f_n) considered, converge uniformly on the images by u' of equicontinuous subsets of E', which follows from Proposition 1.

PROPOSITION 2   Let $H \subset L^1(\mu)$. For H to be relatively compact it is necessary and sufficient that it be weakly relatively compact and relatively compact on the space $\mathscr{M}(\mu)$ (for the topology of convergence on measure on every compact set).

The necessity is clear. For the sufficiency it suffices to show that from every sequence (f_n) extracted from H we can extract a subsequent convergent in L^1. Now we can extract from it a weakly convergent sequence (g_n) (Šmulian) and from this a sequence which converges in measure; since if we suppose M countable at infinity then $\mathscr{M}(\mu)$ is metrisable (therefore every relatively compact sequence of this space admits a convergent subsequence), and in the general case can be reduced to the preceding case noticing that the g_n are zero outside the union of a sequence of compact sets (since each g_n is integrable). The proposition follows from

COROLLARY   Let (f_n) be a sequence in L^1. Then it converges strongly to $f \in L^1$ if and only if it converges weakly and if it converges to f in $\mathscr{M}(u)$ (i.e. in measure on every compact set).

The necessity is obvious. For the sufficiency we notice that the set of f_n being weakly relatively compact in L^1 satisfies conditions a) and b) of Theorem 1. From this, and from the fact that it converges in measure to f on every compact set, we conclude easily that

$$\int |f - f_n|\, d\mu \to 0.$$

COROLLARY 2   In the space l^1 every weakly compact subset is compact, every weakly convergent sequence is convergent.

EXERCISE 1   A subset of L^∞ is relatively compact for $\tau(L^\infty, L^1)$ if and only if it is bounded and relatively compact in $\mathscr{M}$ (space of equivalent classes of measurable scalar functions with convergence in measure on every compact set). A sequence in L^∞ converges for $\tau(L^\infty, L^1)$ if and only if it is bounded and converges in measure on every compact set. (Reduce to the second statement and use Proposition 1.)

EXERCISE 2   Let M be a locally compact space, $\mathscr{M}^1(M)$ the space of bounded measures on M, the dual of $C_0(M)$.

a) Let μ be a positive measure on M. If to every $f \in L^1(\mu)$ we assign the measure $f\mu$ of "density f" with respect to μ we obtain a metric isomorphism from $L^1(\mu)$ into $\mathscr{M}^1(M)$ which respects the natural order structures. We shall write $L^1(\mu) \subset \mathscr{M}^1(M)$. Show that the $L^1(\mu)$

(μ variable in $\mathcal{M}^1(M)$) form an increasingly directed family of closed vector subspaces of $\mathcal{M}^1(M)$ whose union is $\mathcal{M}^1(M)$. For every *sequence* $L^1(\mu_i)$ of such subspaces there exists an $L^1(\mu)$ of the same type that contains them all. (Choose

$$\mu = \sum \frac{1}{2^i} \frac{u_i}{\|u_i\|}$$

and apply the Lebesgue–Nikodym theorem.)

b) Show that we can find a locally compact space M' with a positive measure μ and an isomorphism of ordered Banach spaces from $\mathcal{M}^1(M)$ onto $L^1(\mu)$. (Consider a maximal family $(\bar{\mu}_i)$ of positive measures on M pairwise mutually singular and choose for (M', μ) the sum of the spaces (M_i, μ_i) where M_i is the support of μ_i).

c) A subset of $\mathcal{M}^1(M)$ bounded for the lattice structure is relatively compact for $\sigma(\mathcal{M}^1(M), (\mathcal{M}^1(M))')$.

EXERCISE 3

a) Let H be a weakly compact subset of a space $L^1(\mu)$. Show that we can find a sequence (K_i) of compact subsets of M such that every $f \in H$ vanishes outside $\cup K_i$, therefore that there exists an $f \in L^1$ such that for every $g \in H$ we have $gu \in L^1(fu)$.

b) Conclude, with the notation of Exercise 2, that a weakly compact subset of a space $\mathcal{M}^1(M)$ is contained and is weakly compact in some subspace $L^1\mu$. (Use Exercise 2, b); naturally, by weak topology in $\mathcal{M}^1(M)$ we understand the weak topology defined by the dual of the Banach space $\mathcal{M}^1(M)$ and not the coarser topology $\sigma(\mathcal{M}^1(M), C_0(M))$.

c) Let A be a subset of a space $\mathcal{M}^1(M)$. Show that the following conditions are equivalent:

  1) There exists a positive bounded measure μ on M such that $A \subset L^1(\mu)$;

  2) A is contained in the closed vector space generated by some lattice bounded subset of $(\mathcal{M}^1(M)$;

  3) A is contained in the closed vector space generated by some weakly compact subset of $\mathcal{M}^1(M)$.

Show that the sets A are closed under countable unions. If $\bar{\mu}$ is a positive measure on M then the subset $L^1(\mu)$ of $\mathcal{M}^1(M)$ satisfies the conditions above if and only if μ is countable at infinity.

d) Let μ be a positive measure countable at infinity on the locally compact space M, u a continuous linear mapping from $L^1(\mu)$ into a space $\mathscr{M}^1(M')$ (M' a locally compact space). Show that the image of $L^1(\mu)$ is contained in a space $L^1(\nu)$, where ν is some bounded positive measure on M'. (Use c).)

EXERCISE 4

a) Let K be a compact space, F a quasi-complete LCTVS, u a continuous linear mapping from $C(K)$ into F. For every Hausdorff quotient space $\tilde{K}$ of K we identify $C(\tilde{K})$ to a normed subspace of $C(K)$. Show that u is weakly compact if and only if every metrisable quotient space $\tilde{K}$ of K the restriction of u to $C(\tilde{K})$ is weakly compact. (Use Eberlein's theorem and notice that every *sequence* in $C(K)$ is contained in the space $C(\tilde{K})$ with $\tilde{K}$ some metrisable quotient of K.)

b) For every Hausdorff quotient space $\tilde{K}$ of the compact set K we consider the transpose mapping of the canonical embedding from $C(\tilde{K})$ into $C(K)$, we then obtain a canonical metric homomorphism $\mu \mapsto \phi(\mu)$ from the space $\mathscr{M}(K)$ of measures on K onto the space $\mathscr{M}(\tilde{K})$ of measures on $\tilde{K}$ ($\phi(\mu)$ is known as the *image of the measure* μ by the canonical mapping ϕ from K into $\tilde{K}$). Let A be a subset of $\mathscr{M}(K)$. Show that A is weakly relatively compact (by weak topology on $\mathscr{M}(K)$ we understand the topology $\sigma(\mathscr{M}(K), (\mathscr{M}(K))')$ and not the topology $\sigma(\mathscr{M}(K), C(K)))$, if and only if for every *metric* quotient $\tilde{K}$ of K the canonical image of A in $\mathscr{M}(\tilde{K})$ is weakly relatively compact (Show that A is bounded then reduce to the case where A is disked and closed for $\sigma(\mathscr{M}(K), C(K))$, therefore equal to the image of the unit ball of a dual F' by the transpose of a continuous linear mapping u from $C(K)$ into a Banach space F. Show that u is weakly compact using a)).

EXERCISE 5   Let M be a locally compact space, $\mathscr{M}^1 = \mathscr{M}^1(M)$ the space of bounded measures on M (the dual of $C_0(M)$). Every bounded function f on M measurable for every measure on M (for example a bore function) defines a continuous linear form $\mu \mapsto \int f\,d\mu$ on $\mathscr{M}^1$ of norm $\sup_t |f(t)|$, which permits the identification of the space of these functions with the uniform norm with a normed subspace of the dual of $\mathscr{M}^1$. We call E_1 (resp. E_0) the subspace formed of linear combination of characteristic functions of open sets (resp. of open sets which are the union of a sequence of closed sets). By weak topology in $\mathscr{M}^1$ we understand the topology $\sigma(\mathscr{M}^1, (\mathscr{M}^1)')$.

a) A Cauchy sequence for $\sigma(\mathscr{M}^1, E_1)$ is already weakly convergent. (Use Exercise 2, a) in order to reduce to the case of a sequence in a space $L^1(\mu)$, then use Lemma 1 and Theorem 1.)

b) Let A be a subset of $\mathscr{M}^1$, then A is weakly relatively compact if and only if A is relatively compact for $\sigma(\mathscr{M}^1, E_0)$. (Reduce to the case where M is compact, using the one point compactification, then to the case where M is metrisable by Exercise 4 b). Notice that A is bounded by Chapter 3, Section 3, Exercise 7, c). It suffices then to extract from every sequence (μ_i) in A a weakly convergent subsequence by Eberlein's theorem. Extract first of all a convergent sequence for $\sigma(\mathscr{M}^1, C_0)$ using the fact that the unit ball of $\mathscr{M}^1$ is metrisable and compact for this topology. Then notice that A is also relatively compact for $\sigma(\mathscr{M}^1, \bar{E}_0)$, where $\bar{E}_0$ is the closure of E_0 for the norm topology of $\mathscr{M}^1$ and that $C_0 \subset \bar{E}_0$ so that every sequence in A which converges for $\sigma(\mathscr{M}^1, C_0)$ converges also for $\sigma(\mathscr{M}^1, E_0)$ by compactness, therefore weakly by a)).

c) Let u be a continuous linear mapping from $C_0(M)$ into a quasi-complete LCTVS F. Show that the following conditions are equivalent:

1) u is weakly compact;

2) $u''(E_0) \subset F$;

3) u transforms weak Cauchy sequences into weakly convergent sequences;

4) u transforms every bounded non-decreasing sequence into a weak Cauchy sequence.

(We have immediately 1) $\Rightarrow$ 2) $\Rightarrow$ 3) $\Rightarrow$ 4), by the characterization of weak Cauchy sequences in $C_0(M)$, Part 3, Section 2, Exercise 8; furthermore we easily obtain 4) $\Rightarrow$ 2); finally 2) $\Rightarrow$ 1) by b)).

EXERCISE 6   Let M be a locally compact space, F a quasi-complete LCTVS. We suppose that in F every weak Cauchy sequence is weakly convergent. Show that every continuous linear mapping from $C_0(M)$ into F is weakly compact. (Use Exercise 5, c). Every continuous linear mapping from $C_0(M)$ into a space L^1 or $\mathscr{M}^1(M)$, M' being locally compact, is weakly compact. (Use a) and Theorem 1, Corollary 3), finally, use the fact that $\mathscr{M}^1(M)$ is isomorphic to a space L^1, Exercise 2 b).)

EXERCISE 7   Let M be locally compact with a positive measure μ, E a quasi-complete LCTVS, f a scalarly summable mapping from M into E which defines a natural linear mapping u from E' into $L^1(u)$, ux' being the equivalent class of the function $\langle f(t), x' \rangle$. The transpose of u is a linear mapping from L^∞ into the algebraic dual $\hat{E}$ of E' usually denoted by

$$u'\phi = \int f\phi \, d\mu$$

("weak integral"). Show that $\int f\phi\, d\mu$ is an element of E for every $\phi \in L^\infty$ if and only if this is the case for every ϕ characteristic function of a closed set. (Consider the case where F is a Banach space; then show that u is a *continuous* linear mapping from E' into L^1 using the closed graph theorem, therefore u' is a *continuous* linear mapping from L^∞ into E''. Conclude that it maps $C_0(M)$ into E, then, using Exercise 5, c), Criterion 2), conclude that it is a *weakly compact* linear mapping from $C_0(M)$ into E. Conclude by noticing that the unit ball of $C_0(M)$ is weakly dense in that of L^∞).

EXERCISE 8   Let E be a quasi-complete LCTVS in which every weak Cauchy sequence is weakly convergent (for example, a reflexive space or L^1). Show that in E every scalarly summable sequence is summable. (Use Chapter 2, Section 18, Exercise 3.)

EXERCISE 9   Let M be a locally compact space, let $\mathcal{M}^1 = \mathcal{M}^1(M)$ be the space of bounded measures on M (dual of $C_0 = C_0(M)$), H a subset of $\mathcal{M}^1$. Show that the following conditions are equivalent:

a) H is relatively compact for $\sigma(\mathcal{M}^1, (\mathcal{M}^1)')$.

b) There exists a positive measure μ on M such that $H \subseteq L^1(\mu)$ (when $L^1(\mu)$ is identified with a space of bounded measures as in Exercise 2). Then μ being thus fixed, H as a subset of $L^1(\mu)$ satisfies the conditions a) and b) of the theorem.

c) There exists a positive measure μ on M such that $H \subset L^1(\mu)$. Then H being thus fixed, every bounded sequence (f_n) in $L^\infty(\mu)$ which converges in measure converges uniformly on H.

d) Every sequence in C_0 weakly convergent to 0 (i.e. bounded and converging to zero in every point of M), converges uniformly on H.

e) For every sequence of open sets O_i pairwise disjoint in M we have $\lim \mu(O_i) = 0$ uniformly when μ is in H.

f) H satisfies the two conditions: (i) For every compact $K \subset M$ and every $\varepsilon > 0$, there exists an open set $U \supset K$ such that

$$| \mu | (U \cap \complement K) \leqslant \varepsilon$$

for every $\mu \in H$. (ii) For every $\varepsilon > 0$ there exists a compact $K \subset M$ such that $| \mu | (\complement K) \leqslant \varepsilon$ for every $\mu \in H$.

a) $\Rightarrow$ b) by Exercise 3, b) and Theorem 1; b) $\Rightarrow$ c) by Proposition 1; c) $\Rightarrow$ d) trivially; we show without difficulty d) $\Rightarrow$ e) and e) $\Rightarrow$ f).

It remains to be shown that f) $\Rightarrow$ a), for this consider, using Eberlein's theorem, the case where H is countable and contained in a space

$L^1(\mu)$, using Exercise 2 a). (Then use Theorem 1 where we need only prove condition a); for this we can consider the case where M is compact and we finally prove the conclusion by contradiction, *not* at all trivial.)

EXERCISE 10   Let u be a continuous linear mapping from $C_0(M)$ (M locally compact) into a Banach space E. Show the equivalence of the following conditions:

a) u is weakly compact;

b) there exists a positive measure μ on M such that $u'(E') \subset L^1(\mu)$. Then μ being fixed the following conditions are satisfied:

α) For every $\varepsilon > 0$ there exists $\eta > 0$ such that for every borelian set A with $\mu(A) \leqslant \eta$ we have $\| u''(\phi_A) \| \leqslant \varepsilon$.

β) For every $\varepsilon > 0$ there exists a compact $K \subset M$ such that for every borelian set $A \subset \complement K$ we have $\| u''(\phi_A) \| \leqslant \varepsilon$.

c) There exists a positive measure μ on M such that u can be extended by continuity into a mapping from $L^\infty(\mu)$ into E continuous with respect to $\sigma(L^\infty(\mu) L^1(\mu))$ and $\sigma(E, E')$.

d) u transforms the weakly convergent sequences of $C_0(M)$ into strongly convergent sequences of E.

e) For every sequence (O_i) of open pairwise disjoint subsets of M we have $\lim u''(\phi_{O_i}) = 0$ in E.

f) We have the two properties:

α) For every compact $K \subset M$ and every $\varepsilon > 0$, there exists an open set $U \supset K$ such that for every K' contained in $U \cap \complement K$ we have $\| u''(\phi_{K'}) \| \leqslant \varepsilon$.

β) For every $\varepsilon > 0$ there exists a compact K contained in M such that for every compact K' disjoint from K we have $\| u''(\phi_{K'}) \| \leqslant \varepsilon$.

In the statements of conditions a), c) and f) we have identified as in Exercise 5, the bounded bore functions on M with elements of the bidual of $C_0(M)$. ϕ_A is the characteristic function of A. (Proof: with the exception of c), the equivalence of these conditions is merely the reformulation of the equivalence of the corresponding conditions in Exercise 9. a) $\Rightarrow$ c) follows directly from Exercise 3, b) and c) $\Rightarrow$ a) is trivial because the unit ball of $L^\infty(\mu)$ is compact for the considered weak topology). Extend the equivalence of conditions a, d, e, f (f appropriately modified) to the case where E is a quasi-complete LCTVS.

Compare the results of Exercises 9 and 10 to those of the following exercise (and also of Exercise 5).

EXERCISE 11   Let E be an $\mathscr{F}$ space. A subset H of E' is relatively compact for $\tau(E', E)$ if and only if every weakly convergent sequence in E converges uniformly on H; a linear mapping from E into an LCTVS F transforms weakly compact subsets into compact subsets if and only if it transforms weakly convergent sequences into convergent sequences. (These two statements are equivalent by Chapter 2, Section 18, Theorem 12; the proof is immediate by Šmulian's theorem.)

Let $E = C_0(M)$. A subset of E' is relatively compact for $\tau(E', E)$ if and only if it is so for $\sigma(E', E'')$; a linear mapping u from E into a quasi-complete LCTVS F is weakly compact if and only if it transforms weakly compact subsets into compact subsets. (Use Theorem 1, and Exercise 9. Compare with the next section Proposition 3 and Corollaries, and Theorem 2, and Section 2, Exercise 3, 3), 4), 5).)

EXERCISE 12

1) Let E be a Banach space. Show that the following conditions are equivalent:

a) Every continuous linear mapping from E into a *separable* quasi-complete LCTVS is weakly compact.

b) Every sequence (x_i) of E', which converges weakly to 0, converges to 0 for $\sigma(E', E'')$.

(For a $\Rightarrow$ b we consider the mapping

$$x \longmapsto (\langle x, x_i \rangle)$$

from E into c_0 defined by the given sequence in order to show that the latter is relatively compact for $\sigma(E', E'')$; for b $\Rightarrow$ a it suffices to show that the image of an equicontinuous subset of E' by the transpose u' is relatively $\sigma(E', E'')$-compact, and for this we use Eberlein's theorem noticing that from an equicontinuous sequence of the dual of F' we can extract a weakly convergent subsequence.)

2) Let K be a *Stonian* compact space, i.e. such that $C(K)$ is a complete lattice. Show that $C(K)$ has the properties considered in 1) (We admit that $C(K)$ is a direct factor of the space $l^\infty(K)$. Compare with Chapter 3, Section 3, Exercise 8, c)), this leads us to the case

$$E = l^\infty(K) = C(\hat{K})$$

(K a given index set, $\hat{K}$ its Stone compactification). It suffices to show that the sequence (x_i) is relatively compact for $\sigma(E', E'')$ and for this we apply Criterion c) of Exercise 9 noticing that we can suppose the O_i simultaneously open and closed, K being totally discontinuous. Proceeding as in Chapter 3, Section 3, Exercise 8, a), consider the case where I is the set of integers and $O_i = \{i\}$, then conclude by means of Chapter 3, Section 7, Exercise 2, c).

3) A separable Banach space isomorphic to a quotient of a space $C(K)$, K stonian compact, is reflexive (use 2)). Conclude that if M is an infinite locally compact space with countable base, then $C_0(M)$ is not a direct factor in its bidual. (Notice that when the dual of $E = C_0(M)$ is isomorphic to an L^1 space, Exercise 2, its bidual, is isomorphic to a space L^∞, then to a $C(K)$ with K a stonian compact set. On the other hand, E is separable, and therefore cannot be isomorphic to a quotient of E'' since E is not reflexive.) This generalizes Chapter 3, Section 7, Exercise 2, d).

EXERCISE 13 Let M be a locally compact space with a positive measure μ, E a Hausdorff LCTVS, E' its dual. A mapping f from M into E is said to be *strictly weakly summable* if it is scalarly summable and if for every $\phi \in L^\infty$, the weak integral $\int \phi f \, d\mu$ (which a priori is a linear form on E') is an element of E. Then f defines a linear mapping u_f from L^∞ into E, $u_f(\phi) = \int \phi f \, d\mu$ which is continuous for the weak topology $\sigma(L^\infty, L^1)$ and $\sigma(E', E)$ therefore transforms the unit ball of L^∞ into a *weakly compact* subset of E.

1) If f is a strictly weakly summable mapping from M into E, the mapping u_f transforms the unit ball of L^∞ into a limited subset of E (see Part 3, Section 3, Supplementary Exercise 3, 3). (We must show that every sequence (x'_i) weakly convergent in E' converges uniformly on the subset of E in question or equivalently that the sequence $u'_f x'_i$ is strongly convergent in L^1. By Proposition 2, Corollary 1, it suffices to show that it is a sequence in L^1 which is weakly relatively compact and that it converges at every point.)

2) Conclude that u_f is a compact mapping in each of the following cases: E is separable; E is a reflexive Banach space or more generally the dual F' of an $(\mathscr{F})$ space F with the topology $\tau(F', F)$; E is an L^1 space; f is strongly measurable. (Except for the last case, under each of the preceding conditions, the limited subsets of E are precompact, by Part 3, Section 3, Supplementary Exercise 3, e, and Supplementary Exercise 5; in the last case, reduce to the case where E is a Banach space —using Part 3, Section 2, Exercise 3—then the case where M is countable at infinity, finally the case of E separable noticing that we can assume that f is defined on a separable subspace of E.) Note: We do not know whether uf is compact in all cases. We can easily reduce the problem to the case where E is the Banach space l^∞ and M is compact proceeding as above, and thus to the case where M is metrisable or

even where M is the segment $(0, 1)$ with Lebesgue measure, and we can finally assume f to be bounded using Section 2 Supplementary Exercise 4, 3).

3) Let f be a scalarly summable mapping from M into the complete space E. Show that f is strictly weakly summable if and only if the mapping from E' into $L^1(\mu)$ defined by f transforms equicontinuous subsets into weakly relatively compact subsets of L^1, and that we can find a total subset H of the weak dual L^∞ of L^1 such that $\int \phi f \, d\mu \in E$ for every $\phi \in H$. (Notice that u_f is then a mapping from L^∞ into the algebraic dual E'^* of E' continuous for $\tau(L^\infty, L^1)$ and for the equitopology of equi-continuous convergence on E' mapping the dense vector subspace generated by H into the complete, therefore closed subspace E of E'^*.) Note: we do not know whether the second condition stated above is superfluous. We can reduce this question to the case where E is a Banach space, M compact and finally (by Section 2, Supplementary Exercise 4, 3) f scalarly essentially bounded.

2 Application of the Dunford–Pettis criterion

PROPOSITION 3 *Let E be an LCTVS, $\mathfrak{S}$ a set of bounded subsets of E, $\mathfrak{S}'$ the set of subsets of E' which are disked, equicontinuous and compact for $\sigma(E', E'')$. The following assumptions on $(E, \mathfrak{S})$ are equivalent:*

a) *Every continuous linear mapping u from E into a Hausdorff LCTVS F which transforms bounded subsets into weakly relatively compact subsets, transforms the $A \in \mathfrak{S}$ into relatively compact subsets.*

b) *The $A \in \mathfrak{S}$ are precompact for the $\mathfrak{S}'$-topology.*

c) *The $A' \in \mathfrak{S}'$ are precompact for the $\mathfrak{S}$-topology.*

We point out immediately that in the definition of $\mathfrak{S}'$ it was not necessary to suppose the A''s disked (by Krein's theorem, Part 3, Section 3).

The proof is standard: a) $\Rightarrow$ b), since if we take F to be the completion of E for $\mathfrak{S}'$-convergence and u the identity mapping of E (with its given topology) into F, u satisfies the conditions of a) by Chapter 2, Section 18, Theorem 13 (the transpose u' transforms an equicontinuous subset of F' into a set contained in an element of $\mathfrak{S}'$, thence equicontinuous and $\sigma(E', E'')$ compact), therefore transforms the A into relatively compact subsets of F, i.e. the $A \in \mathfrak{S}$ are precompact for $\mathfrak{S}'$-convergence. b) $\Rightarrow$ c) by Chapter 2, Section 18, Theorem 12 applied to the preceding mapping u from E into F. Finally c) $\Rightarrow$ a) by the same

Q

theorem applied to u. (We find that u transforms the $A \in \mathfrak{S}$ into *precompact* subsets of F, but since they are also weakly relatively compact by the hypothesis on u, A being bounded, they will be even relatively compact—Chapter 2, Section 18, Proposition 37, Corollary 1.)

COROLLARY 1   *Let E be a quasi-barrelled space, E' its strong dual, suppose that E' satisfies the conditions of Proposition 3 when $\mathfrak{S}'$ stands for the set of $\sigma(E', E'')$-compact subsets of E'. Then E satisfies the analogous condition (relative to the set of weakly compact subsets of E).*

We use the notations of Proposition 3. The hypothesis on E' means also (when we choose the statement b) of the proposition, which we apply to $(E', \mathfrak{S}')$) that the $A' \in \mathfrak{S}'$ are precompact for the topology of uniform convergence on the set $\mathfrak{S}''$ of subsets A'' of E'' which are equicontinuous and $\sigma(E'', E''')$ compact. But an $A \in \mathfrak{S}$, i.e. a weakly compact subset of E, clearly belongs also to $\mathfrak{S}''$, (since E being quasi-barrelled is a topological vector subspace of E'' strong), therefore a fortiori the $A' \in \mathfrak{S}'$ are precompact for $\mathfrak{S}$-convergence, which is merely Condition b) of Proposition 3 for $(E, \mathfrak{S})$.

COROLLARY 2   *Let E be an $(\mathscr{F})$ space, $\mathfrak{S}$ the set of weakly compact subsets of E. For the conditions of Proposition 3 to be satisfied it is (necessary and) sufficient that every sequence weakly convergent to 0 in E, converges uniformly on every $A' \in \mathfrak{S}'$.*

We shall use the Condition b) of Proposition 3. The necessity is trivial. For the sufficiency we must show that every weakly compact disked subset A of E is precompact for $\mathfrak{S}'$-convergence. Now (Šmulian's theorem) we are able to extract from the given sequence a weakly convergent subsequence which by the hypothesis will also converge in the $\mathfrak{S}$-topology.

COROLLARY 3   *Let E be an LCTVS, $\mathfrak{S}$ the set of subsets of E which are weak Cauchy sequences. Then the conditions of Proposition 3 are equivalent to the following:*

a') *Every continuous linear mapping u from E into a Hausdorff LCTVS F, which transforms bounded subsets into weakly relatively compact subsets, transforms weak Cauchy sequences into convergent sequences.*

b') *Every weak Cauchy sequence in E is also a Cauchy sequence for the $\mathfrak{S}'$-convergence.*

If these conditions are satisfied, and if E is of type $(\mathscr{F})$, then the conditions of Proposition 3 are still satisfied if $\mathfrak{S}$ is the set of weakly compact subsets of E.

It is in fact immediate that a′) resp. b′) are equivalent ways of expressing a) resp. b). Finally, the last part of the corollary follows from Corollary 2.

THEOREM 2 *Let E be a space $L^1(\mu)$ on $C_0(M)$, u a mapping from E into a Hausdorff LCTVS F which transforms bounded subsets into weakly relatively compact subsets. Then u transforms weakly compact subsets into relatively compact subsets and weak Cauchy sequences into convergent sequences.*

The first assertion means that E satisfies the conditions of Proposition 3 when $\mathfrak{S}$ is the set of weakly compact subsets of E. Furthermore, since the dual L^∞ of a space L^1 is isomorphic to a space $C(K)$ (Kakutani, Stone), this assertion on L^1 spaces will be already established if we show it for the $C_0(M)$ spaces (by Proposition 3, Corollary 1). Then, the second assertion will also be proved for the case $E = L^1$ since in L^1 a weak Cauchy sequence is weakly convergent (Theorem 1, Corollary 3). We are thus reduced to the case $E = C_0(M)$. Furthermore, by Proposition 3, Corollary 3, it suffices to prove the second assertion of the theorem which can also be stated: Let (f_i) be a weak Cauchy sequence in $E = C_0(M)$, A' a subset $\sigma(E', E'')$ compact in E'. Show that (f_i, μ) converges uniformly when μ runs through A'. If this is not the case we could find $\varepsilon > 0$, extract from A' a sequence (μ_j) and find a strictly increasing sequence (i_j) of indices such that

$$| \langle f_{i_{j+1}} - f_{i_j}, \mu_j \rangle | \geqslant \varepsilon.$$

This leads us to the case where A is a weakly relatively compact sequence. But we know that a bounded positive measure μ then exists such that all the μ_i are absolutely continuous with respect to μ

for example: $\mu = \sum \dfrac{1}{2^i} \dfrac{u_i}{\| u_i \|}$

so that the μ_i can be identified with elements of $L^1(\mu)$. Since $L^1(\mu)$ is a normed subspace of the space $\mathcal{M}^1(M) = E'$ of bounded measures on E, which is closed (since it is complete) therefore also closed for $\sigma(E', E'')$, the sequence A' is also weakly relatively compact in $L^1(\mu)$. On the other hand, (f_i) being a weak Cauchy sequence, it is bounded in $C_0(M)$, and furthermore converges in each point, and a fortiori converges in measure on every compact set with respect to μ. By Proposition 1, it follows that $\langle f_i, \mu \rangle$ converges uniformly when μ runs through A'; this ends the proof.

COROLLARY 1 *Let E, F, G be LCTVS, u a continuous bilinear mapping from $E \times F$ into G. We suppose that E is a space isomorphic to a space*

L^1 or to a space $C_0(M)$. Then for every weakly compact subset A of E and B of F the restriction of u to $A \times B$ is continuous for the product of the weak topologies on $A \times B$, and the weak topology on G (in particular, $u(A \times B)$ is a weakly compact subset of G).

We can reduce this to the case where u is a continuous bilinear form. By Chapter 2, Section 18, Theorem 12, it suffices to show that the linear mapping v from E into F' defined by u transforms the weakly compact subsets A into subsets of F' which are compact for $\tau(E', E)$. Now v is a weakly compact mapping when we equip F' with the topology $\tau(F', F)$, then the conclusion follows from Theorem 2. Also:

COROLLARY 2 *Under the conditions of Corollary 1, let (x_i) be a weak Cauchy sequence in E, (y_i) a weak Cauchy sequence in F, then $(u(x_i, y_i))$ is a weak Cauchy sequence in G.*

Reduce to the case where u is a bilinear form, which we interpret as a linear mapping from F into E'.)

The most important application of Theorem 2 is

THEOREM 3 (DUNFORD–PETTIS–PHILIPPS) *Let M be a locally compact space with a positive measure μ, E a Banach space, u a weakly compact linear mapping from $L^1(\mu)$ into E. Then there exists a measurable mapping f from M into E, such that*

$$\| f(t) \| \leqslant \| u \|$$

for every t, and that

$$(1) \qquad u\phi = \int \phi(t) f(t) \, d\mu(t).$$

for every $\phi \in L^1(\mu)$. (Notice that the product ϕf will be automatically integrable which gives meaning to the formula—see Bourbaki, Integration.) This f is unique modulo the locally negligible functions.

The uniqueness is immediate and well known in a more general context. It follows that it suffices to prove the theorem when M is compact: for then, to every compact set K we assign a class F_K of bounded and measurable functions f on K with values in E, with

$$\| f(t) \| \leqslant \| u \|$$

for almost every $t \in K$, such that (1) will be satisfied for the ϕ which vanish in $\complement K$; the uniqueness shows that $K \subset K'$ implies that F_K is the restriction of $F_{K'}$; by a well-known lemma due to Godement, specially conceived for this sort of situation, there exists a mapping f from

M into E which induces on each K the class F_K. Then we verify immediately that this mapping f satisfies the desired conditions ($\| f(t) \| \leqslant \| u \|$ locally almost everywhere only, and a modification of f on a locally negligible set allows the same inequality everywhere).

Suppose M compact. Then $L^\infty \subset L^1$, and the unit ball of L^∞ is a weakly compact subset of L^1 (for example by Theorem 1, Corollary 2), hence (Theorem 2) it is transformed into a relatively compact subset of E. Let E_1 be the closed vector subspace generated by the latter; it is a *separable* Banach space and since L^∞ is dense in L^1 and mapped into E_1, u maps also L^1 into E_1. This reduces the problem to the case where E is also separable, which we now suppose. Let A be the closure of the image of the unit ball of L^1. It is a weakly compact subset of E, and since E' is weakly separable (E being separable) A will be metrisable for its weak topology. A is also the unit ball of the dual of the Banach space F obtained (passing to the quotient and then to the completion) from the space E' with the semi-norm gauge of A^0, and with this identification the two weak topologies are identical. Since A is a metrisable compact set for its weak topology, it follows that F is separable. Thus u becomes a linear mapping of norm 1 from L^1 into the dual of the separable Banach space F. From the Dunford–Pettis theorem in its classical form (as in Bourbaki), u is defined by a scalarly mapping f from M into the unit ball of F' by the integral (1) (but where the integral is a weak integral in F'). A fortiori f is a scalarly integrable mapping from M into E, and formula (1) is valid as a weak integral in E. It remains to point out that in fact f is even measurable; it is well known that for a mapping with range a *separable* Banach space scalarly measurable implies measurable. Finally the inequality $\| f(t) \| \leqslant \| u \|$ is implied by the more precise result $f(M) \subset A$ (true without supposing M compact nor E separable). As a first interesting application see Exercises 8, 9. Theorem 3 plays an important part in the theory of topological tensor products.

EXERCISE 1 Show that we obtain a condition equivalent to the conditions of Proposition 3 if we state b) supposing that F is a Banach space or even the Banach space l^∞ (in the latter case, use Part 3, Section 2, Exercise 3).

EXERCISE 2 We say that an LCTVS E is a *DP* space (Dunford–Pettis) if it satisfies the conditions of Proposition 3, $\mathfrak{S}$ being the set of its weakly compact subsets.

1) If E is *DP* so is every direct factor of E. The analogous result is false for the subspaces and quotients (recall that *every* Banach space is

isomorphic to a subspace of a space $C(K)$ and to a quotient space of a space L^1, see Chapter 1, Section 14, Exercise 1 and Chapter 2, Section 17, Proposition 30, Corollary).

2) A reflexive DP Banach space is finite dimensional. (Show that its unit ball is precompact.)

3) A reflexive Banach space of infinite dimension is not isomorphic to a direct factor of a space L^1 or of a space $C_0(M)$ (use 1) and 2)). Using the remarks at the end of 1), deduce examples of vector subspaces in $C(K)$ or L^1 which have no topological supplements.

Exercise 3

1) Let u be a mapping from a uniform space E into a Hausdorff topological space F. If u transforms Cauchy sequences into convergent sequences, it transforms precompact and metrisable subsets into relatively compact subsets of F. (Show that u can be extended by continuity to the completion of the precompact metrisable subset in question.) Corollary: Let u be a mapping from a uniform space E into another, F, transforming Cauchy sequences into Cauchy sequences, then u is uniformly continuous on every metrisable precompact subset, which is transformed by u into a precompact subset. (Consider the case where F is Hausdorff and complete, then, from the preceding argument u can be extended by continuity to the completion of the precompact subset under consideration, then the conclusion follows.)

2) Particular cases of 1). Let u be a linear mapping from a Hausdorff LCTVS F transforming weak Cauchy sequences into convergent sequences, then u transforms the weakly metrisable bounded subsets into relatively compact subsets.

3) In particular, if in F the weak Cauchy sequences converge weakly, then every weakly continuous linear mapping from E into F transforms weakly metrisable bounded subsets into weakly relatively compact subsets. In particular (choosing $E = F$, u the identity mapping of F) every weakly metrisable bounded subset of F is weakly relatively compact. If the strong dual of E is separable, every weakly continuous linear mapping from E into F transforms bounded subsets into weakly relatively compact subsets.

Exercise 4

1) A weakly metrisable bounded subset of a space L^1 is weakly relatively compact (use Theorem 1, Corollary 3 and Exercise 3, 3)).

2) Let E be a $C_0(M)$ space of infinite dimension. Show that there exists in E weak Cauchy sequences not weakly convergent (a fortiori

therefore weakly metrisable bounded subsets not weakly relatively compact), and weakly convergent sequences not convergent. (Proceed either by direct construction, considering the case of M metrisable by passage to the quotient, or by noticing that if this were not true the identity mapping of E would be weakly compact by Section 1, Exercise 5 c, Criterion 3) and Exercise 10, Criterion d); then E would be reflexive which necessitates that E is finite dimensional.

3) Conclude from 1) and 2) that a space $C_0(M)$ of infinite dimension is not isomorphic to a topological vector subspace of a space $L^1(\mu)$ and neither is a space L^1 of infinite dimension isomorphic to a quotient of a space $C_0(M)$ (the second statement is equivalent to the first one by duality; we can also prove the latter first, using Section 1, Exercise 6, b)).

EXERCISE 5

1) Let E and F be Banach spaces, E isomorphic to a space L^1 or $C_0(M)$, u a weakly compact bilinear form on $E \times F$, A a weakly compact subset of E, B the unit ball of F. Show that the restriction of u to $A \times B$ is continuous for the product of the weak topologies. (Consider u as a weakly compact linear mapping from E into the Banach space F' strong and apply Theorem 2.) We can also replace the hypothesis on A by the following: A is bounded and weakly metrisable. (Use Exercise 3, 3).) There is a variant supposing that F is an LCTVS; what is the hypothesis to be made on u?

2) Let E, F be Banach spaces each isomorphic to some space $C_0(M)$. Let u be a continuous bilinear form on $E \times F$, show that if A is a weakly compact subset or a bounded and weakly metrisable subset of E, B the unit ball of F, the restriction of u to $A \times B$ is continuous for the product of the weak topologies. (Use 1) and Section 1, Exercise 6 b).)

EXERCISE 6

1) Let μ be a bounded positive measure on a locally compact space M. Let E be a vector subspace of $L^\infty(\mu)$ closed in an L^p space with $1 \leqslant p < \infty$. Show that E is finite dimensional. (Using the closed graph theorem show that on E the topology induced by L^∞ or L^p is the same if $p = 1$, which allows reduction to reduce the case $p > 1$; show then that the unit ball of E induced by L^p is precompact, applying Theorem 2 to the identity mapping from L^∞ into the reflexive space L^p.)

2) Let μ be a positive measure on a space M. Let $1 \leqslant p < \infty$, let E be a closed vector subspace of L^p and $f_0 \in L^p, f_0 \geqslant 0$, such that every

$f \in L^p$ is bounded above by some multiple of f_0. Show that E is finite dimensional. (Introduce the bounded measure of density $|f_0|^p$ with respect to μ, and reduce to 1).)

EXERCISE 7   Prove Theorem 2 for the case where E is an $(\mathscr{F})$ space. (Reduce to the case of a Banach space, by Chapter 4, Part 2, Section 2, Theorem 1, Corollary 2.)

EXERCISE 8   Let M be a locally compact space with a positive measure μ, f a scalarly measurable mapping from M into a Banach space E, g a strongly measurable mapping from M into E, scalarly locally almost everywhere equal to f.

1) Show that f defines a weakly compact linear mapping from L^1 into E and apply Theorem 3. In particular, every scalarly measurable and locally bounded mapping from M into a *reflexive* Banach space E is scalarly locally equal almost everywhere to a strongly measurable mapping from M into E. (Note: A deeper study using supplementary exercise 4, 3) below would show that this result is still valid without supposing the given mapping locally bounded, and if the space E is a reflexive $(\mathscr{F})$ space.)

2) According to general definitions a mapping f from M into E is *weakly measurable* if for every compact $K \subset M$ and every $\varepsilon > 0$ there exists a compact $K' \subset K$ such that $\mu(K \cap \complement K') \leqslant \varepsilon$ and such that the restriction of f to K' is weakly continuous. (This implies that f is scalarly measurable but the converse is false.) Show that a weakly measurable mapping from M into E is already strongly measurable. (Reduce to the case where f is weakly continuous and M compact. Then apply 1) noticing that we even have $f(t) = g(t)$ almost everywhere and not only scalarly a.e. For the latter point reduce to the case where g is itself continuous.)

3) Particular case of 2): let f be a weakly continuous mapping from a locally compact space M into a Banach space E, then f is measurable for every measure μ on M.

EXERCISE 9   Let M, N be compact spaces, f a numerical function on $M \times N$. For every $s \in M$ let $F(s)$ be the function on N given by $F(s)(t) = f(s, t)$.

1) The mapping $s \mapsto F(s)$ is weakly continuous from M into the Banach space $C(N)$ if and only if f is bounded and continuous in each variable. (The necessity is trivial; for the sufficiency one proves that

$F(M)$ is a weakly compact subset of $C(N)$, using Part 3, Section 2, Theorem 3.)

2) Suppose that the mapping f is continuous in each variable. Show that for every positive measure μ on M and every $\varepsilon > 0$, there exists a compact set $K' \subset K$ such that $\mu(K \cap \complement K') \leqslant \varepsilon$, and such that the restriction of f to $K' \times N$ is continuous. (Reduce to the case where f takes its values in the segment $[0, 1]$ therefore satisfies the conditions of 1); it suffices then to apply Exercise 10, 3).)

3) f being as in 2), show that f is measurable for every positive measure μ on $M \times N$. (Let ρ be the image of μ by the projection of $M \times N$ on M, then apply the result of 2) to f and ρ.) Show that 2) and 3) are still valid if f takes its values in a separable metrisable space P and if in 3) we suppose only M and N locally compact. (Imbed P into the product of a sequence of segments, thus reduce to the case of a function with values in a segment; in 3), if M and N are only locally compact reduction to the compact case is immediate.)

4) Suppose the numerical function f on the product of compact M and N to be bounded and continuous in each variable. For every measure μ on M and ν on N f is integrable for $\mu \otimes \nu$ by 3); set

$$\mu_f(\mu, \nu) = \int\limits_{M \times N} f \, d\mu \, d\nu$$

Show that we thus obtain a weakly separately continuous bilinear form on the product $\mathscr{M}(M) \times \mathscr{M}(N)$ of duals of $C(M)$ and $C(N)$, the corresponding mapping from $\mathscr{M}(M)$ into $C(N)$ being

$$\mu \mapsto \int\limits_{M} F \, d\mu$$

(weak integral in $C(N)$). (Use 1).) Conversely, every weakly separately continuous bilinear form u on $\mathscr{M}(M) \times \mathscr{M}(N)$ is defined by an f as above uniquely determined by

$$f(s, t) = u(\varepsilon_s, \varepsilon_t)$$

(ε_s being the measure $+1$ at s). We have

$$\| u \| = \sup_{s,t} | f(s, t) |.$$

EXERCISE 10   Let U be an open set of $\mathbf{C}^n$, $f(z_1, \ldots, z_n)$, a function defined on U locally summable or locally bounded and holomorphic in each variable separately. If f is locally bounded, show that if f is defined on the product of two open sets $U \subset \mathbf{C}^p$ and $V \subset \mathbf{C}^q$, locally bounded and separately holomorphic in U and V then f is holomorphic. Since it is measurable by Exercise 11, 3), it will be locally summable and we examine this case. Consider now the distribution T defined by f and show that it satisfies

$$\frac{\partial}{\partial z_i} T = 0$$

by a holomorphic function g, a.e. equal to f. Show finally $f = g$.

Supplementary exercises

These exercises are not tied directly to the text but use the techniques therein.

EXERCISE 1

1) Let $E = L^1(\mu)$ constructed on a measure countable at infinity. Let H be a convex subset of the dual $E' = L^\infty(\mu)$. Show that the following conditions are equivalent:

   a) H is weakly closed;

   b) H is closed with respect to weakly convergent sequences;

   c) H is closed for the bounded sequences which converge, a.e. a) ⇒ b) trivially; b) ⇒ c) since the sequences considered in c) are weakly convergent by Lebesgue's theorem. In order to show c) ⇒ a) reduce to the case where H is bounded by the Banach–Dieudonné theorem (Chapter 4, Part 2, Section 3, Theorem 2), then to the case where μ is bounded, replacing μ by an equivalent bounded measure therefore to the case $L^\infty \subset L^1$. (Show that H is then a *closed* convex subset of L^1, therefore weakly closed in L^1, thence weakly compact in L^1, since it is already weakly relatively compact, and it follows that H is also weakly compact in L^∞.)

2) Let E be a *separable* space, A a convex subset of E'. Then A is weakly closed if and only if it is weakly closed for the sequences. (Use the Banach–Dieudonné theorem.) (This exercise could have been given in Chapter 4, Part 2, Section 3.)

3) Let f be a mapping from a Hausdorff locally compact space M into a locally convex space E. Show that the subspace H of E' formed of x' such that $\langle f(t), x' \rangle$ is measurable is a vector subspace which is closed for the weakly convergent *sequences*. Conclude that if E is a separable ($\mathscr{F}$) space or if E is an L^1 space constructed on a measure countable at infinity, then f is scalarly measurable if there exists a total

subset of E' made up of forms x' such that $\langle f(t), x' \rangle$ is measurable. (Note: this result becomes false if we choose for example $E = l^\infty$ or $E = l^1(I)$ with I not countable.)

4) Let E be a $C(K)$ space constructed on a compact Stonian space (see Exercise 12) for example l^∞. Let H be a convex subset of E'. Then H is closed for the weakly convergent sequences if and only if H is *strongly* closed. (Use Exercise 12, 2).) In particular, if dim $E = \infty$ we can find in E' a hyperplane closed for the weakly convergent sequences which is not weakly closed, therefore we can find a linear form continuous for the weakly convergent sequences which is not weakly continuous. (In fact, E is not reflexive.)

5) Show that 1) is false if we do not suppose μ countable at infinity. (Take the space E of 4) and recall that a Banach space is always isomorphic to a quotient of a space $l^1(I)$ on an appropriate index set I.)

EXERCISE 2

1) Let $E = L^1(\mu)$ constructed on a measure μ countable at infinity. Let u be a linear form on E'; show that the following conditions are equivalent:

    a) u is weakly continuous;

    b) u is continuous for the sequences that tend weakly to 0;

    c) u is continuous for the sequences bounded in L^∞ which tend to 0 almost everywhere. (Use Exercise 1, 1).)

2) Let $E = L^1(\mu)$ with μ on a locally compact space M. Show that we can find a locally compact space $\hat{M}$ a topological sum of a family of compact spaces (M_i), and a measure ρ on $\hat{M}$ of support $\hat{M}$ such that $L^1(\rho)$ is isomorphic as an ordered Banach space, to $L^1(\mu)$. We can suppose I reduced to one element, or not countable; the power of I is well determined and called the *order at infinity* of the measure μ. (Proceed as in Section 1, Exercise 2, b).)

3) With the notations of 2), let $p =$ power of $I =$ order at infinity of μ. Show that the following conditions are equivalent:

    a) Every linear form on the dual of $L^1(\mu)$ continuous for the weakly convergent sequences, is weakly continuous;

    b) Every continuous form on the dual $l^\infty(I)$ of $l^1(I)$, continuous for the weakly convergent sequences, is weakly continuous.

    c) Every positive linear form on l^∞ zero on c_0, continuous for the weakly convergent sequences, is zero.

(a) $\Rightarrow$ b) is easy, b) $\Rightarrow$ a) is easy by 1) and 2); for c) $\Rightarrow$ b) consider the restriction v of u to c_0 as an element of l^1, i.e. a weakly continuous linear form on l^∞, and show that $u - v$ is zero which reduces b) to the case of a form u zero on c_0. Interpreting $l^\infty(I)$ as the space $C(\hat{I})$ ($\hat{I}$ the Stone compactification of I) show that if the measure μ on I is a linear form on $l^\infty(I)$ continuous for the weakly convergent sequences, so is $|\mu|$ (which leads us to the case $\mu \geqslant 0$). For this, interpreting the continuous linear forms on $l^\infty(I)$ as the bounded "additive set functions" (defined on the family of all subsets of E), the forms which are continuous for the weakly convergent sequences become the "*completely additive*" set functions and our assertion means that the "absolute variation" of a completely additive set function is still completely additive, which is well known although condemned by Bourbaki).

4) In the language of set functions, the hypothesis considered in 3), which deals only with the cardinal p, means that there does not exist a completely additive positive set function defined on the family of all subsets of I for which the measure of points is 0 and which is not equal to 0. We say that such a cardinal is a cardinal of *measure* 0. If there exists a cardinal which is not of measure 0, there exists a first one p_0, (and all the following cardinals do not have measure 0). Show that the sum of a family of cardinals of measure 0 is of measure 0 if the power of the index family is also of measure 0; thus the following cardinal is of measure 0, therefore p_0 if it exists is a limit cardinal. A cardinal is *inaccessible* if it is a limit cardinal and if for every family of strictly inferior cardinals with an index set of strictly inferior power its sum is strictly inferior. Therefore p_0 would be an inaccessible cardinal. It seems plausible that we can add without contradictions, to the axioms of set theory, the non-existence of inaccessible cardinals, and a fortiori that every cardinal has measure zero (compare with Chapter 4, Part 1, Section 6, Exercise 3, g).

EXERCISE 3

1) Let μ be a positive measure on a locally compact space; consider $E = L^1(\mu)$ as an ordered Banach space. Show that if H is a closed vector subspace which is a sublattice, there exists a uniquely determined projection from E onto H transforming positive elements into positive elements and conserving the norm of the positive elements; this projection has norm 1 if $H \neq 0$. (Reduce to the case where M is a compact K, then, by consideration of the Kakutani space K, i.e. the space K such that $L^\infty(\mu)$ is isomorphic to $C(K)$, in which case the natural

mapping $C(K) \to L^\infty(\mu)$ is an isomorphism onto. Considering

$$H \cap L^\infty = H \cap C(K),$$

show that we can find a mapping ϕ from K onto a Hausdorff quotient K' of K in such a way that denoting μ' the image of μ by ϕ (see Section 1, Exercise 4, b), H is identical to the set of classes of functions $f' \circ \phi$, where $f' \in L^1(\mu')$. Identifying $L^1(\mu)$ with the space of measures on K of base μ, show that if the measure ρ on K is in $L^1(\mu)$ then its image in K' is in $L^1(\mu')$. The desired projection is obtained by composing the mappings

$$\rho \mapsto \phi(\rho) \quad \text{and} \quad f' \mapsto f' \circ \phi)$$

2) Show that every separable subspace of $L^1(\mu)$ is contained in a separable direct factor. (Consider the closed vector sub-lattice generated by the given subspace.) Note: this TVS property of L^1 spaces is not shared by the spaces L^∞ (as usual!).

EXERCISE 4   Let μ be a positive measure countable at infinity on the locally compact space M. Let $\hat{M}(\mu)$ be the set of classes of real functions finite or not which are measurable on M (which has therefore a largest element $+\infty$ and a smallest element $-\infty$).

1) Let H be an increasingly directed subset of $\hat{M}(\mu)$; show that the filter of increasing sections on H tends to a limit for convergence in measure on every compact set and that this limit is the upper bound of H in $\hat{M}(\mu)$. We can find a sequence in H which converges a.e. to this upper bound. (Examine the case where μ is bounded, therefore $L^\infty \subset L^1$, then examine the classical analogous properties, for the directed sets bounded in L^1, identifying $\hat{M}$ with the set of measurable functions which lie between 0 and 1, by virtue of a strictly increasing homomorphism ϕ from $[-\infty, \infty]$ onto $[0, 1]$.)

2) Let H be a set of real finite measurable functions on M (not classes of functions) such that for every $t \in M$ we have $\sup_{t \in H} f(t) < +\infty$. Show that there exists a real, finite, measurable function g on M such that $f \in H$ implies $f(t) \leqslant g(t)$ a.e. (Show with the aid of the last part of 1), that the sup in $\hat{M}$ of the set of classes of $f \in H$ is the class of a function which is finite a.e.

3) Let f be a scalarly measurable function from M into an LCTVS E, let A be a weakly bounded subset of E'. Show that we can find a real measurable finite function g on M such that for every $x' \in A$ we have

$$| \langle f(t), x' \rangle | \leqslant g(t) \text{ a.e.}$$

We say that f is *scalarly essentially bounded* if for $x' \in E'$, the class of the function $\langle f(t), x' \rangle$ is in L^∞. Deduce from the above that if E is a Banach space (or more generally an ($\mathscr{F}$) space), for every compact $K \subset M$ and every $\varepsilon > 0$ there exists a compact $K' \subset K$ such that

$$\mu(K \cap \complement K') \leqslant \varepsilon$$

and such that the restriction of f to K' is scalarly bounded.

4) Let E be a Banach space. Suppose that the pair (μ, E) to be such that the Dunford–Pettis theorem characterizing the continuous linear mappings from L^1 into E', or equivalently, from E into L^∞, is valid. (This is the case if E is separable, or if M is a Kakutani space, etc.) Let u be a continuous linear mapping from E into $\mathscr{M}(\mu)$. The mapping u can be defined by means of a scalarly measurable function f from M into E' by the usual formula $ux = $ class of the function $\langle x, f(t) \rangle$, if and only if the image of the unit ball of E in $\mathscr{M}(\mu)$ is bounded for its lattice structure (for the necessity use 3); for the sufficiency, reduce to the case where u maps the unit ball of E into the unit ball of L^∞, by composition of the given mapping with some multiplicative mapping.

5) Suppose the measure μ non-discrete (for example the Lebesque measure on $[0, 1]$). Show that for $1 \leqslant p < \infty$ there exist continuous linear mappings from $E = L^1(\mu)$ into $L^p(\mu)$ which cannot be obtained by a mapping from M into $E' = L^\infty(\mu)$ i.e. by a kernel function which is measurable on $M \times M$. (Replace E by l^1 noticing that l^1 is isomorphic to a quotient of L^1. Then show that there exist in L^p bounded sequences not bounded for the lattice structure in $\mathscr{M}(\mu)$ and which define mappings from l^1 of the desired type.)

EXERCISE 5   Let M be a locally compact space with a positive measure μ, f a strictly weakly summable mapping from M into a space E of type ($\mathscr{F}$) (see Section 1, Exercise 13). We suppose that in E every limited subset is precompact and that every separable subspace of E is contained in a separable direct factor (this condition is verified in particular if E is a space L^1 constructed on an arbitrary measure by Exercise 3, 2) and Part 3, Section 3, Supplementary Exercise 5). Then there exists a *measurable* mapping g from M into E, scalarly locally equal to f a.e. (by Section 1, Exercise 13, 1), the mapping u_f from L^∞ into E defined by f is compact; let F be a separable direct factor of E containing the image of L^∞ by u_f, let p be a continuous projection from E onto F, we let $g = p \circ f$. Then g is measurable since g is a mapping into a *separable* ($\mathscr{F}$) space F. Observe that $u_f = u_g$, which means that f is equal to g scalarly locally a.e.

EXERCISE 6

1) Let M be a locally compact space with a positive measure μ, E a

Hausdorff locally convex space such that every linear form on E' which is continuous for the equicontinuous weakly convergent sequences is continuous (for example, E separable and complete since then the restrictions of the forms considered to the equicontinuous subsets are weakly continuous, since these subsets are weakly metrisable; or, E a space L^1 constructed on a measure μ' whose order at infinity has measure zero, see Exercise 2.) Let f be a scalarly measurable bounded mapping from M into E and let $\phi \in L^1(\mu)$. Show that the weak integral $\int \phi f \, d\mu$ is in E. (It suffices to show that it is a continuous form for the weakly convergent equicontinuous sequences and for this we apply Lebesque's theorem.)

2) If we suppose furthermore E metrisable and that, in E, every weak Cauchy sequence is weakly convergent, then we have the stronger result: every scalarly summable mapping from M into E is strictly weakly summable. (By Section 1, Exercise 13, 3), it suffices to show that the mapping from E' into L^1 defined by f transforms equicontinuous subsets into weakly relatively compact subsets of L^1 and that $\int \phi f \, d\mu \in E$ for every continuous function with compact support ϕ on M; using Eberloin's theorem reduce to the case where M is countable at infinity, then using the second hypothesis on E and Exercise 4, 3) reduce to the case where M is compact and f scalarly essentially bounded on M. One is then back to 1).)

EXERCISE 7 Let M, M' be locally compact spaces with positive measures μ, μ'; let $E = L^1(\mu')$. We suppose that the order at infinity of μ' is a cardinal of measure zero or that for every compact $K \subset M$, the power of K is a cardinal of measure zero (see Exercise 2). Let f be a scalarly measurable mapping from M into E; show that there exists a *measurable* mapping g from M into E which is equal to f scalarly almost everywhere; if f is weakly summable g is strictly weakly summable (Section 1, Exercise 13). (Show that we can suppose M compact using Godement's lemma in the first case and proceeding as in the preceding exercise 2) in the second case. This allows us to assume that we are in case where the order at infinity of μ' is of measure zero, replacing if necessary μ by

$$\mu'' = \sup_{t \in M} \inf\left(1, |f(t)|\right) \mu'.$$

The second assertion results from Exercise 6. For the first assertion, reduce to the case where f is scalarly essentially bounded using Exercise 4, 3); then f is scalarly summable, therefore strictly weakly summable, and we can apply Exercise 5).